SpringerBriefs in Electrical and Computer Engineering

Control, Automation and Robotics

SpringerBriefs in Control, Automation and Robotics presents concise summaries of theoretical research and practical applications. Featuring compact, authored volumes of 50 to 125 pages, the series covers a range of research, report and instructional content. Typical topics might include:

- a timely report of state-of-the art analytical techniques;
- a bridge between new research results published in journal articles and a contextual literature review;
- a novel development in control theory or state-of-the-art development in robotics;
- an in-depth case study or application example;
- a presentation of core concepts that students must understand in order to make independent contributions; or
- a summation/expansion of material presented at a recent workshop, symposium or keynote address.

SpringerBriefs in Control, Automation and Robotics allows authors to present their ideas and readers to absorb them with minimal time investment, and are published as part of Springer's e-Book collection, with millions of users worldwide. In addition, Briefs are available for individual print and electronic purchase.

Springer Briefs in a nutshell

- 50–125 published pages, including all tables, figures, and references;
- softcover binding;
- publication within 9–12 weeks after acceptance of complete manuscript;
- copyright is retained by author;
- authored titles only – no contributed titles; and
- versions in print, eBook, and MyCopy.

Indexed by Engineering Index.

Publishing Ethics: Researchers should conduct their research from research proposal to publication in line with best practices and codes of conduct of relevant professional bodies and/or national and international regulatory bodies. For more details on individual ethics matters please see: https://www.springer.com/gp/authors-editors/journal-author/journal-author-helpdesk/publishing-ethics/14214

Rifat Sipahi · Silviu-Iulian Niculescu ·
Fatihcan M. Atay

Deterministic Car-Following Traffic Models

Delay Effects and Linear Stability

Springer

Rifat Sipahi
Department of Mechanical and Industrial Engineering
Northeastern University
Boston, MA, USA

Silviu-Iulian Niculescu
Laboratoire des Signaux et Systèmes
University Paris-Saclay, CNRS, CentraleSupélec, INRIA
Gif-sur-Yvette, France

Fatihcan M. Atay
Department of Mathematics
Bilkent University
Ankara, Türkiye

ISSN 2191-8112 ISSN 2191-8120 (electronic)
SpringerBriefs in Electrical and Computer Engineering
ISSN 2192-6786 ISSN 2192-6794 (electronic)
SpringerBriefs in Control, Automation and Robotics
ISBN 978-3-031-58163-2 ISBN 978-3-031-58164-9 (eBook)
https://doi.org/10.1007/978-3-031-58164-9

Mathematics Subject Classification: 34K05, 93C23, 93D09, 93D20, 34K20, 34K35, 37N35, 93C05, 93C10

This Springer imprint is published by the registered company Springer Nature Switzerland AG
The registered company address is: Gewerbestrasse 11, 6330 Cham, Switzerland

To (the memory of) our parents, Nuzhet & Selahattin Sipahi, Eugenia & Mihai Niculescu, and Saadet & Adnan Atay

Preface

Traffic dynamics attract extensive interest of many researchers from a variety of scientific disciplines including physics, mathematics, and engineering. Investigations on traffic dynamics go back at least to the 1930s, to the work of Greenshields [1]. The reason for the vast interest in this research area stemmed from the fact that the volume of traffic has been rapidly increasing and exceeding the capacities of highways, as Greenberg stated six decades ago [2]. Following these works, many studies have been devoted to understand vehicular flow based on various mathematical models, up until today [3–8].

The importance of traffic research was reconfirmed in the 1990s, especially in rapidly growing cities in the US, Germany, the UK, France, and Japan, as cities with large populations suffered from increased highway commute times and traffic congestion, calling for higher capacity infrastructures and better traffic flow management. Indeed, optimizing the capacity to manage traffic load has become critical. A Texas A&M Transportation Institute report observes that wasted time and fuel due to waiting in congested traffic cost the US economy billions of dollars [9].

Some undesirable effects of traffic can be minimized by carefully managing traffic flow. For instance, city traffic lights can be controlled to regulate vehicle cross flow and provide real-time traffic re-routing information for the drivers to avoid congestion, or vehicles with adaptive cruise control systems can regulate their motion to prevent aggressive accelerations and decelerations, which are known to cause undesirable effects downstream of traffic flow. In order to achieve a well-managed traffic flow, one of the main goals would be to understand the underlying mechanisms of traffic flow dynamics, which requires creating realistic *mathematical models*.

Traffic flow models can be categorized in different ways, including *stochastic* versus *deterministic*, *continuous-time* versus *discrete-time*, and *macroscopic* versus *microscopic*. For example, stochastic discrete-time models can be used to study traffic flow [10], macroscopic models can be used to investigate global metrics like traffic flow rate and density, and microscopic models can be used to describe the contribution of each vehicle to traffic flow using continuous-time differential equations [11]. In each case, one would adopt an appropriate modeling scheme depending on the conditions at hand and the parameters of interest, which may include multiple versus single-lane traffic, car-following dynamics, lane departure and overtaking a vehicle, human driver models, autonomous vehicle models, merging traffic on ramps into highway traffic, existence of traffic lights, and consideration of obstacles in the lane.

One particular class of traffic flow models widely studied is *car-following dynamics*, where vehicles follow each other in a single lane without overtaking and lane changes, which corresponds mainly to driving patterns on highways. Study of this problem is important because highways are the arteries of transportation, carrying significant amounts of traffic load and transferring goods and people rapidly over long distances. The impact of traffic jams and accidents on highways is particularly significant as they can involve a large amount of stranded people, wasted work hours, economic losses, and emissions to the environment. According to a news article published in the UK [12], vehicles idling in traffic in the UK, France, and Germany due to gridlock release about 2000 kilotons of carbon dioxide, corresponding to a carbon cost of £100 million per year that could only be offset by planting trees every year in an area equivalent to half the size of Greater London.

Several factors play a role in shaping car-following dynamics, such as the way people drive their vehicles, the ability of vehicles to accelerate/decelerate, the speed at which vehicles travel, and the safety headway between vehicles. One key question is whether the headway between vehicles can be kept at a safe level, and how headway is affected downstream of vehicles when a perturbation is triggered upstream. In some cases, the headway can be perturbed from upstream in such a way that downstream vehicles may need to reduce their speeds significantly, or even stop completely, to prevent a crash. Technically, headway perturbations and their propagation can be studied using two mathematical notions, namely *stability* and *slinky effects*. In the current context, stability refers to the ability of the flow dynamics to dissipate energy against impulsive disturbances, such that the headway between vehicles converges back to the equilibrium state after perturbations. Slinky effects, on the other hand, pertain to cyclic perturbations, causing the headway between consecutive vehicles downstream to grow in amplitude. Both loss of stability (i.e., instability) and slinky effects are undesirable in traffic flow dynamics, as they may cause "stop-and-go" waves responsible for traffic jams and increased emissions.

A crucial detail is then how drivers react to headway changes. This naturally varies from driver to driver, in addition to being dependent on a number of other factors. For example, aggressive drivers may introduce high accelerations/decelerations, thereby perturbing the headway propagation undesirably downstream, while calmer drivers may induce very different effects. A critical physiological parameter pertaining to driver influence is *human reaction delays*, including delays in perception, decision-making, execution of decisions, and muscular response. For instance, when the driver sees the brake lights of the preceding vehicle, it may take up to two seconds for them to push the brake pedal. On the one hand, such a delayed action may cause a crash if the speed is high and the headway is not large enough. On the other hand, even if a crash is avoided, the effects of perturbations on the headway will propagate downstream, disrupting the traffic flow.

Due to human reaction delays, traffic flow dynamics is inherently a *time-delay system*, and the governing equations are *delay differential equations* [13–15]. These are generally more difficult to study than ordinary differential equations, mainly because the presence of delays creates an "infinite-dimensional" system. Specifically, the characteristic equation of a linear delay differential equation has infinitely many roots, whose locations on the complex plane need to be studied for understanding the stability of the system. Finite-dimensional tools are usually not applicable and it is necessary to draw upon results from the delay systems literature [16, 17] to investigate stability, slinky effects, and beyond.

The goal of this brief is to provide a compact overview of deterministic car-following models and the mathematical tools for analyzing the stability of traffic flow dynamics in the presence of time-delays. The models considered here are *microscopic flow models*, formulated as continuous-time deterministic delay differential equations, with a focus on the linear stability and slinky effects. Therefore, firstly, a knowledge of the linearization of nonlinear models and the basics of linear systems are desirable to comfortably follow the text. Moreover, the stability analysis presented takes an eigenvalue approach; therefore, further concepts from linear systems, transfer functions, complex analysis, and sensitivity analysis of a system's poles will be used in the development. Nevertheless, we have aimed to make the material self-contained as much as possible.

This volume starts with an overview of fundamental mathematical models of traffic flow from the literature, including a brief survey of microscopic car-following models, experimental studies, and software and simulation tools. Chapter 2 discusses the *source of human reaction delays*, mainly arising from human physiology, as well as factors affecting reaction delays, which can be modeled mathematically by discrete or distributed delay terms. We then present a list of mathematical models in which such reaction delays are considered, including discussions of tools utilized, types of analysis performed, and major findings. This is followed in Chap. 3 by a mathematically oriented presentation, where we define the notions of stability and slinky effects on dynamical systems and consider the effects of constant delays. In Chap. 3, computational tools for the study of time-delay systems are also highlighted, and some of the inherent challenges in the qualitative theory of delay differential equations are pointed out. In Chaps. 4–6, the presented mathematical tools are implemented on various traffic flow models for the study of stability and slinky effects. Through these examples, we also present the ways in which delays and human aggressiveness can affect linear traffic flow characteristics, show how ring and chain formation of vehicles give rise to different dynamical properties with interpretations through graphs and decomposition ideas, and demonstrate how some of these discussions can be implemented in an adaptive cruise control setting.

Boston, USA
Gif-sur-Yvette, France
Ankara, Türkiye
February 2024

Rifat Sipahi
Silviu-Iulian Niculescu
Fatihcan M. Atay

References

1. B.D. Greenshields, J. Rowland Bibbins, W.S. Channing, H.H. Miller, A study of traffic capacity, High. Res. Board **14**, 44–477 (1935)
2. H. Greenberg, An analysis of traffic flow, Operat. Res. **7**, 78–85 (1959)
3. I. Prigogine, *Theory of Traffic Flow*, R. Herman, Amsterdam (ed.) (Elsevier, 1961)
4. F.A. Haight, *Mathematical Theories of Traffic Flow* (Academic Press, New York, 1963)
5. G.B. Whitham, *Linear and Nonlinear Waves* (Wiley, New York, 1974)
6. D. Chowdhury, L. Santen, A. Schadschneider, Statistical physics of vehicular traffic and some related systems, Phy. Rep. **329**, 199–329 (2000)
7. B. S. Kerner, *Introduction to Modern Traffic Flow Theory and Control: The Long Road to Three-Phase Traffic Theory* (Springer, 2009)
8. M. Treiber, A. Kesting, *Traffic Flow Dynamics. Data, Models and Simulation* (Springer, Berlin, Heidelberg, 2013)
9. The Texas A&M Transportation Institute Report, Associated Press, February 5, 2013, published at http://www.nydailynews.com/autos/stuck-traffic-car-commuters-wasted-121b-2011-article-1.1255719. Accessed 9 Jun 2023
10. B.-H. Wang, Y. R. Kwong, P. M. Hui, Statistical mechanical approach to cellular automaton models of highway traffic flow, Phy. A **254**, 122–134 (1998)
11. D. Helbing, Traffic and related self-driven many-particle systems, Rev. Mod. Phy. **73**, 1067–1141 (2001)
12. UK news, December 18, 2013, published at http://www.fleetpoint.org/fleet-industry-news/news-by-date/traffic-congestion-costs-uk-households-4-4billion-inrix-study-finds/12810/ Accessed 9 Jun 2023
13. G. Stépán, *Retarded Dynamical Systems: Stability and Characteristic Function* (Longman Scientific & Technical, New York, Co-publisher: John Wiley & Sons Inc., USA, 1989)
14. J. K. Hale, S. M. Verduyn Lunel, *Introduction to Functional Differential Equations* (Springer-Verlag, New York, 1993)
15. R. Bellman, K. L. Cooke, *Differential-Difference Equations* (Academic Press, New York, 1963)
16. W. Michiels, S.-I. Niculescu, *Stability and Stabilization of Time-Delay Systems. An Eigenvalue Based Approach* (SIAM, Philadelphia, 2007)
17. R. Sipahi, S.-I. Niculescu, C.T. Abdallah, W. Michiels, K. Gu, Stability and stabilization of systems with time delay, limitations and opportunities. IEEE Con. Sys. Mag. **31**(1), 38–65 (2011)

Acknowledgments

The authors of this volume have been collaborating for almost two decades on time-delay systems, through meetings that took place at the Max Planck Institute for Mathematics in the Sciences in Leipzig (former affiliation of Dr. Atay), at Supélec (now CentraleSupélec) in Gif-sur-Yvette, at Northeastern University in Boston, as well as through participation and organization of numerous conferences and workshops, and, of course, online meetings. The authors gratefully acknowledge the financial support of Chateaubriand Bourse of the French government, which supported the postdoctoral work of Dr. Sipahi at HeuDiaSyC (CNRS) Université de Technologie de Compiègne during 2005–2006; French CNRS and CentraleSupelec at Gif-sur-Yvette through some international research networks projects (DELSYS, 2012–2015 and SPA-DISCO, 2018–2021); the Max Planck Institute for Mathematics in the Sciences in Leipzig; and Northeastern University, Boston. Dr. Atay took the department chair position at the Department of Mathematics at Bilkent University, Ankara, Turkey, whose support is also gratefully acknowledged. Dr. Sipahi would also like to thank Dr. Min Hyong Koh, Dr. Adrian Ramirez, Dr. Duo Wang, and Mr. Haonan Fan for their careful reading of and recommendations in Chaps. 1–2.

The authors would like to thank the Editor Oliver Jackson and the production team for their precious time and meticulous handling of this Brief.

Boston, MA, USA
Gif-sur-Yvette, France
Ankara, Türkiye
February 2024

Rifat Sipahi
Silviu-Iulian Niculescu
Fatihcan M. Atay

Contents

Chapter 1
Traffic Flow Modeling

In this chapter, we present a classification of traffic flow models aimed at capturing various physical phenomena and representing different traffic flow conditions, followed by an in-depth discussion of microscopic models. Example models are provided that account for individual vehicle dynamics from both deterministic and stochastic perspectives. Experimental studies and examples of simulation tools are surveyed. The chapter concludes with notes and relevant references.

1.1 Classification of Traffic Flow Models

In general, one can classify traffic flow models as being either deterministic or stochastic in nature. Most of the *models* discussed in this chapter are *deterministic*, while some insight into stochastic effects will also be covered. In the deterministic case, one encounters a rich volume of literature, mainly with two distinct directions: one in which the focus is on parameters such as flow density and flow rate associated with *macroscopic* properties of traffic, and another one that focuses on the driver behavior and dynamic effects of vehicles pertaining to *microscopic* features of traffic flow. These models, to a certain degree, can also accommodate various parameters of interest, such as flow diffusion rate, propagation of headway between consecutive vehicles, effects of individual drivers and driving strategies, road and weather conditions, and physical constraints of the environment including speed limit posts. Although this list is not exhaustive, it already shows how intricately the models can be developed.

In the sequel, we summarize some fundamental models by following a standard classification path established in the literature. Along with the models, various physical interpretations and application areas are discussed.

R. Sipahi et al., *Deterministic Car-Following Traffic Models*,
SpringerBriefs in Control, Automation and Robotics,
https://doi.org/10.1007/978-3-031-58164-9_1

1.1.1 Macroscopic Models

These models capture the macroscopic features of traffic flow dynamics, ideally those that can be measured, so that the models can be studied in connection with experiments. In these models, two key variables are the *traffic density* and the *traffic flow rate* (flux), denoted here, respectively, by c and J. These variables depend continuously on the spatial coordinate x describing the physical location of traffic, and time t. That is, one has $J = J(x,t)$ and $c = c(x,t)$, and the flow dynamics can be expressed using partial differential equations. Leveraging the hydrodynamic theory of fluids, macroscopic flow models can be derived from the fundamental equations of fluid dynamics and are thus named after "fluid dynamics theories" [1, 2]. As an example, considering traffic as a one-dimensional compressible fluid, the continuity equation can be expressed as

$$\frac{\partial c(x,t)}{\partial t} + \frac{\partial J(x,t)}{\partial x} = \mathscr{S}_0 - \mathscr{S}_1, \tag{1.1}$$

where $\mathscr{S}_0$ and $\mathscr{S}_1$ denote, respectively, the sources and sinks, e.g., on-ramp traffic and outgoing, e.g., off-ramp traffic flow. In the absence of on/off ramps, i.e., when the sources and sinks are set to zero, some fundamental macroscopic traffic flow models can be developed, as explained next.[1]

The Lighthill and Whitham model [5], also known as *LW model*, proposes a reasonable approximation by suggesting that the flux $J(x,t)$ is a function of the density $c(x,t)$. With this, one can rewrite the continuity equation as

$$\frac{\partial c(x,t)}{\partial t} + \frac{\partial J(x,t)}{\partial x} = 0, \tag{1.2}$$

$$J(x,t) = j(c(x,t)), \tag{1.3}$$

where j describes the aforementioned function and the flux is given specifically by $J = c(x,t)\vartheta(c(x,t))$, where the velocity ϑ is assumed to depend only on density. Under this setting, one obtains the nonlinear partial differential equation

$$\frac{\partial c(x,t)}{\partial t} + \frac{\partial c(x,t)}{\partial x}\left(\vartheta(x,t) + c(x,t)\frac{d\vartheta}{dc}\right) = \frac{\partial c(x,t)}{\partial t} + \vartheta_g\frac{\partial c(x,t)}{\partial x} = 0, \tag{1.4}$$

where $\vartheta_g = dJ/dc$ represents the speed of density waves relative to the traffic flow and moving in the opposite direction of the flow.

A further modification is possible in (1.3) by adding a diffusion term in the definition of flux as

$$J(x,t) = j(c(x,t)) - D\frac{\partial c(x,t)}{\partial x}, \tag{1.5}$$

[1] See, for example, [3, 4] where authors utilize macroscopic models with ramps.

where D is a positive constant. This modification suggests that a positive density gradient in the traffic results in lower flux, and vice versa. Indeed, one would expect the drivers to reduce their speeds as they merge into a more congested region in traffic. A more detailed discussion on this approach can be found in [2], which we summarize as follows.

Using (1.5) instead of (1.3) in Eq. (1.2), one arrives at the continuity equation

$$\frac{\partial c(x,t)}{\partial t} + \frac{dj(c)}{dc}\frac{\partial c(x,t)}{\partial x} = D\frac{\partial^2 c(x,t)}{\partial x^2}. \tag{1.6}$$

Greenshields [6] further writes the flux as a function of density:

$$J = \vartheta_{\max} c(1-c), \tag{1.7}$$

where $\vartheta_{\max}$ is a constant. With this modification, (1.6) becomes

$$\frac{\partial c}{\partial t} + \vartheta_{\max}\frac{\partial c}{\partial x} - 2\vartheta_{\max} c\frac{\partial c}{\partial x} = D\frac{\partial^2 c}{\partial x^2}, \tag{1.8}$$

where we have dropped the arguments for conciseness. Via the linear transformation $x = \vartheta_{\max} t' - x'$ and $t = t'$, the so-called *deterministic Burgers equation* [7] is derived from (1.8), which reads

$$\frac{\partial c}{\partial t'} + 2\vartheta_{\max} c\frac{\partial c}{\partial x'} = D\frac{\partial^2 c}{\partial x'^2}. \tag{1.9}$$

Equation (1.9) is encountered also in the fluid dynamics literature [8], in Stokes' First Problem, where the fluid sits on top of an infinite plate and an impulsive input to the plate causes the plate to move at a constant speed. Equation (1.9) can then be used to study the effects of plate motion on the fluid where D is analogous to fluid viscosity, and c corresponds to the velocity field of the fluid.

Interestingly, the Burgers equation can be further manipulated. For this, consider the following generic Burgers equation:

$$\frac{\partial c}{\partial t'} = \frac{\partial}{\partial y}\left(D\frac{\partial c}{\partial y} - \frac{c^2}{2}\right), \tag{1.10}$$

and define a nonlinear transformation as introduced by Cole–Hopf [9, 10] in a domain "y"

$$\eta(y,t') = e^{-\frac{1}{2D}\int c dy}, \tag{1.11}$$

which can be alternatively expressed as

$$c(y,t') = -\frac{2D}{\eta}\frac{\partial \eta}{\partial y}. \tag{1.12}$$

Substituting the above expression into the Burgers equation (1.10), and after some simple manipulations, one arrives at the linear heat equation

$$\frac{\partial \eta}{\partial t'} = D\frac{\partial^2 \eta}{\partial y^2}, \tag{1.13}$$

which can be analytically solved for a given set of boundary and initial conditions. It then becomes possible to transform the solution accordingly to the solution of the corresponding Burgers equation, and ultimately to the Greenshields model.

In general, macroscopic models are studied based on both continuity and momentum equations. Referring to [2], one can consider the following "Navier–Stokes-like momentum equation"

$$c\left(\frac{\partial \vartheta}{\partial t} + \vartheta\frac{\partial \vartheta}{\partial x}\right) = -D\frac{\partial c}{\partial x} + \mu\frac{\partial^2 \vartheta}{\partial x^2} + \frac{c}{t_d}\left(\vartheta_{\text{safe}}(c) - \vartheta\right), \tag{1.14}$$

where D is the diffusion constant, μ represents the constant viscosity, and t_d is a constant relaxation time associated with the dynamics of velocity ϑ as it relaxes toward a desired safe velocity $\vartheta_{\text{safe}}(c)$ depending on the density.

Further remarks: Macroscopic models have been utilized to study a variety of phenomena in traffic flows [11–14]. One of the advantages of macroscopic models is that they concentrate on global metrics that are directly related to desirable/undesirable traffic flow characteristics, e.g., density, flux, and synchronized flow. This explains the considerable amount of effort that has been dedicated to studying traffic flow from a macroscopic perspective.

When developing macroscopic models, it is critical to have a good understanding of the description of traffic flow, especially when interpreting fluid-dynamical concepts. For example, the definition of viscosity for traffic flow should be carefully drawn and interpreted; see [2, p. 220] for discussions from a macroscopic point of view and [15, p. 135] from a microscopic point of view. Moreover, one also needs expertise to carefully set up the models with proper assumptions under the scenarios considered, and solving the coupled nonlinear partial differential equations using numerical methods.

In the Navier–Stokes-like momentum Eq. (1.14), one may consider the relaxation time t_d to be regulated by density; see [16] where the resulting model is reported to capture *stop-and-go*-type *congestion* patterns in traffic. In [17] a Boltzmann-like approach is presented. In [18] the authors start with the continuity Eq. (1.1) without sinks and sources and use the momentum Eq. (1.14) to investigate the evolution of the shape of traffic jams; comparing their findings to Lightwill–Whitham (LW) model as well as experimental results.

1.1.2 Microscopic Models

Microscopic models take into account the dynamics of individual vehicles in traffic. Thus, every vehicle is treated as a particle with its respective dynamics, and the interactions between the vehicles become the focus, i.e., the traffic flow is a set of particles interacting with each other. Some of the relevant parameters in microscopic models are the instantaneous velocities of individual vehicles, the instantaneous headway between consecutive vehicles, the response characteristics of drivers, safe following distance/desired headway, the desired velocity of drivers, the speed limit of the highway under consideration, the physical structure of the highway, including the existence of single/multiple lane(s), with or without on/off ramps, and traffic regulations and policies.

Next, a classification of mathematical models based on the above parameters is discussed. We first introduce *car-following models*, which are widely studied in the literature. Next, we introduce models with discrete-time and space, called *cellular automaton models* or *particle hopping models*. Finally, we discuss the link between the kinetic theory of gases and the traffic flow.

1.1.2.1 Car-Following Models

These types of models are developed based on different driving strategies of the drivers as well as different configurations of vehicles; see Figs. 1.1 and 1.2. The fundamental idea is that the drivers follow at least one car ahead of them to maintain safe driving conditions.[2] A driver's decision is formed by a certain "stimulus", which can be in the form of a velocity error or headway with the preceding vehicle, which elicits a certain "response", in the form of an acceleration or deceleration of the vehicle. For instance, when the driver senses that the headway is smaller than it should be, they will apply the brakes. Since in this case the traffic flow is modeled in the form of "follow the leader", these types of models are mostly used for studying single-lane traffic flow, assuming that every driver is satisfied with maintaining a safe velocity and headway in their respective lane, without any desire to change lanes. (See also [20, 21] for multi-lane models.)

One of the earlier microscopic models is given by Reuschel in 1950 (see discussions in [1] as well as by Pipes [23]). This model is a set of linear differential equations given by

$$\ddot{x}_k(t) = \alpha(\dot{x}_{k+1}(t) - \dot{x}_k(t)), \quad k = 1, \ldots, n, \tag{1.15}$$

where $\ddot{x}_k(t)$ and $\dot{x}_k(t)$ represent, respectively, the acceleration and the speed of the kth vehicle, and α is a constant that can be interpreted as the sensitivity and response characteristics of driver k against the stimulus provided by the velocity differences.

[2] Depending on the model, a vehicle may receive stimulus from multiple vehicles such as found in connected vehicle systems [19].

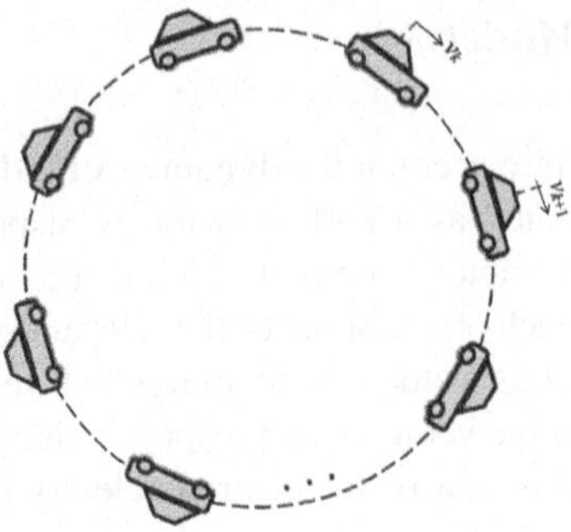

Fig. 1.1 Ring configuration of vehicles

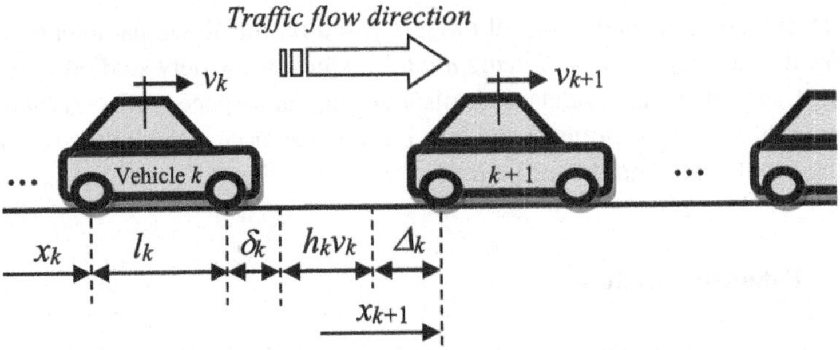

Fig. 1.2 Chain configuration of vehicles, inspired by [22]: Each vehicle with dimension l_k is positioned at $x_k(t)$ and travels at a speed of $v_k(t)$. The spacing between consecutive vehicles is desired to be $h_k v_k(t) + \Delta_k$, where Δ_k is the minimum headway required between the vehicles and $h_k \geq 0$ is known as the time headway. Whenever $h_k > 0$, the additional headway is considered to be proportional to the speed of the vehicle, making the dynamics a constant time-headway problem; otherwise it is said to be a constant-headway problem. The deviation from the desired spacing is denoted by δ_k

Bando in 1995 proposed the Optimal Velocity Model [24] given by

$$\ddot{x}_k(t) = \alpha(V(\Delta x_k(t)) - \dot{x}_k(t)). \tag{1.16}$$

Here $\Delta x_k(t) = x_{k+1}(t) - x_k(t)$ is the headway between consecutive vehicles and $V(\cdot)$ is, in general, a nonlinear function that represents the *optimal velocity*, or the desired velocity that the driver k wishes to reach, for a given headway $\Delta x_k(t)$. In view of their general structure, one can call such models *Optimal Velocity Models (OVM)*.

Inspecting the two models (1.15) and (1.16), we arrive at the analogy that the velocity of the vehicle in the front $\dot{x}_{k+1}(t)$ in (1.15) corresponds to a sophisticated nonlinear optimal velocity function in Bando's model. In light of experimental results obtained from the highways in Japan, Bando in [25] builds upon the above model by specifying an explicit definition of the *optimal velocity function V* given by,

$$V(\Delta x) = 16.8(\tanh(0.086(\Delta x - 25)) + 0.913). \tag{1.17}$$

See also [26–28] for different forms of V. The optimal velocity function V can be selected based on the following considerations:

- In general, V is set to zero for headway less than a certain threshold value selected as the safe stopping distance without causing collision.
- V should be a monotone increasing function of the headway. This is because when the headway decreases, V should decrease as well, and when the headway starts to increase due to preceding vehicle's acceleration, then the driver following this vehicle will usually tend to accelerate their vehicle seeking a higher equilibrium speed.
- V should have an upper bound, i.e., $V(\Delta x) \leq V_{max}$, which can represent the legal speed limit of the highway or maximum velocity to which a vehicle can reach without sacrificing safe driving conditions.

Heterogeneous vehicles are considered in several studies, using OVM as the basic model. For instance, the work in [29] considers cars and trucks traveling in traffic on circuitry and straight roads. The effects of vehicle heterogeneity are investigated inspired by Bando's model (1.16). Criteria for linear stability are studied and propagation of small disturbances downstream of traffic flow is analyzed. Moreover, starting from Bando's nonlinear model, effects on traffic flow are assessed based on the ratio of low-bandwidth trucks to high-bandwidth vehicles in traffic. Enhancement in traffic flow by reducing jam formations has been one of the major foci in the literature. For instance, in [26] the authors utilize the discrete-time version of the model (1.16) along with a piecewise linear optimal velocity function to assess linear stability and to study the design parameters that avoid the growth of position and velocity amplitudes downstream of traffic. They also propose a type of delayed-feedback controller for each vehicle in a decentralized manner in order to suppress jam formations. Along similar lines, the authors in [27] propose to take advantage of automated vehicles to reduce traffic jam formations. To this end, traffic is envisioned to be made of several groups, each of which contains only a few vehicles. Specifically, a leader vehicle operated manually is followed by two automated vehicles in a group, each type of vehicle respecting the model (1.16) but having its own optimal velocity functions. The authors investigate the effects of different numbers of vehicles per group using linear stability theory and find out that with the presence of automated vehicles a better flow rate can be achieved in high-density traffic conditions, compared to having only manually operated vehicles in groups.

In [30], the author studies an optimal velocity model (OVM) to derive necessary and sufficient conditions of linear stability for steady-state flow of traffic in a circular path. The study in [31] provides an excellent survey of various models, including models to predict loads in transportation network, not limited to microscopic models. In [28], the authors describe a number of vehicles on a circular path with different driver characteristics and optimal velocity functions. They primarily investigate how different levels of density in traffic affect jam formations, by carrying out a linear stability analysis as well as studying bifurcations. In [32] the authors consider a time delay in OVM with piecewise linear optimal velocity functions and develop an analytical approach to explain, using asymptotic trajectories, the transition between congested flow and free flow, also known as homogeneous flow. In [33] an improved stabilization scheme is proposed by modifying Bando's model, where a driver follows the two nearest vehicles ahead to make acceleration/deceleration decisions. This idea enables a level of anticipation as the information of headway further upstream can be used to make better driving decisions. Similar ideas were also proposed in [34, 35]. In [34], the driver receives stimulus from more than one vehicle ahead while in [35] the driver receives stimulus from its immediate neighbors in the front and at the back:

$$\ddot{x}_k(t) = \alpha(V_1(\Delta x_k(t)) + V_2(\Delta x_{k-1}(t)) - \dot{x}_k(t)), \tag{1.18}$$

where V_1 and V_2 are optimal velocity functions for two different headways. Along these lines, in [36], the authors propose the idea of following multiple vehicles, i.e., using multiple headway measurements from upstream to inform the desired speed via an optimal velocity function $V(\Delta x_k(t), \Delta x_{k+1}(t), \Delta x_{k+2}(t), \ldots)$. They prove by linear stability analysis that following more than one preceding vehicle improves the stability of traffic flow dynamics. For their analysis they investigate a discretized form of OVM, on which they also perform nonlinear analysis and numerical simulations to study kink anti-kink soliton solutions.

Starting from the OVM model (1.16), Nagatani investigates in [37] the effects of fluctuations in the leading car dynamics on the formation of traffic jams in downstream traffic. In this work, patterns of congestion, localized clusters, and oscillatory flow are identified and shown to exist under certain fluctuation features of the leading car in the traffic flow. Some modified optimal velocity functions are proposed in [38] and their effects on multi-phases and phase separations in traffic flow patterns are studied. In [39], it is shown that different forms of optimal velocity functions have a strong influence on the arising multi-phase traffic.

As highlighted in [15], in certain cases, OVM may produce acceleration and deceleration levels much larger than what is observed in traffic. For example, accelerations may reach $12\,\mathrm{m/s^2}$ while empirical data suggest accelerations up to $4\,\mathrm{m/s^2}$. A remedy to this, as discussed in [15], is the so-called T3 model [40]:

$$\ddot{x}_k(t) = \frac{1 + b_1\dot{x}_k + b_2\Delta x_k + b_3\dot{x}_k\Delta x_k + b_4\dot{x}_{k+1} + b_5\dot{x}_k\dot{x}_{k+1}}{c_0 + c_1\dot{x}_k + c_2\Delta x_k + c_3\dot{x}_k\Delta x_k + c_4\dot{x}_{k+1} + c_5\dot{x}_k\dot{x}_{k+1}}, \tag{1.19}$$

where b_ℓ and c_ℓ are constant model parameters, and headway terms include vehicle lengths. With its eleven independent parameters, the T3 model is shown to satisfactorily capture real-world traffic data, albeit at the cost of identifying eleven parameters versus only five parameters in OVM. Furthermore, considering models of behavioral changes and pedestrian dynamics [41, 42], the study in [15] starts with social force models, which sum *generalized forces* to determine behavioral changes. In traffic, these behavioral changes correspond to velocity changes. These models are hence called *generalized force models* (GFM), where the acceleration of a vehicle is determined by relaxing the velocity to a desired value and its interaction with neighboring vehicles via a nonlinear function f_k. In the general form, one can write

$$\ddot{x}_k(t) = \alpha(\dot{x}_{k,\text{desired}}(t) - \dot{x}_k(t)) + f_k(x_k(t), x_{k+1}(t), \dot{x}_k(t), \dot{x}_{k+1}(t)), \tag{1.20}$$

where $\dot{x}_{k,\text{desired}}(t)$ represents the desired velocity of the kth vehicle and f_k is a state-dependent function, which can be chosen in a variety of ways so long as it guarantees safe driving conditions.[3] In [15], the authors provide specific details including physical intuition for modeling f_k to represent vehicle dynamics. To this end, the proposed GFM model is given by[4]

$$\ddot{x}_k(t) = \alpha(V(\Delta x_k(t)) - \dot{x}_k(t)) + \lambda \Delta \dot{x}_k \Theta(-\Delta \dot{x}_k(t)), \tag{1.21}$$

where $\Delta \dot{x}_k(t) = \dot{x}_{k+1}(t) - \dot{x}_k(t)$, Θ is the Heaviside function, and λ is selected as a function of headway $x_{k+1}(t) - x_k(t)$, where a safe headway is determined by how fast the vehicle is moving. For further details, including a different form of optimal velocity function V, we refer to [15]. Ultimately, one arrives at a GFM model that requires identifying seven free parameters, which has been shown to fit realistic data satisfactorily.

Using the model (1.21), the work in [43] shows that it is possible to reach good agreement with empirical data when α is identified to be two times smaller[5] than that identified for the model (1.16). In their study, a further modification is proposed over (1.21), which reads

$$\ddot{x}_k(t) = \alpha(V(\Delta x_k(t)) - \vartheta_m) + \alpha(\vartheta_m - \dot{x}_k) + \lambda \Theta(-\Delta \dot{x}_k) \Delta \dot{x}_k + \lambda \Theta(\Delta \dot{x}_k) \Delta \dot{x}_k,$$

where a new parameter ϑ_m is defined as the maximum velocity. This model is called *full velocity difference model* (FVDM) since it takes into account both the positive and negative velocity differences between consecutive vehicles. Simulations indicate that this new model is able to capture more reasonably the acceleration levels observed in the real world [43, Fig. 2]. Furthermore, it is found that the transition from homogeneous flow to congestion can be observed, and the criterion for preserving the stability of linearized traffic flow dynamics can be obtained analytically.

Analogous to GFM in the sense of mathematical representation, one can also refer to Treiber's *Intelligent Driver Model* (IDM) [44], which takes into account different acceleration and deceleration characteristics of the vehicles:

$$\ddot{x}_k(t) = \alpha \left(1 - \left(\frac{\dot{x}_k(t)}{\vartheta_{0,k}} \right)^{\delta} - \left(\frac{s^*}{\Delta x_k(t)} \right)^2 \right), \tag{1.22}$$

[3] See [15, p. 136 Eq. (14)] for an example of how to judiciously make such a selection.

[4] The polarity of relative speed in this brief is opposite of that defined in the cited study.

[5] In the cited study a particular optimal velocity function is proposed.

where $\vartheta_{0,k}$ is the desired speed of vehicle k; s^* is the desired minimum gap, which is a function of vehicle velocity $\dot{x}_k(t)$ and headway rate $\Delta\dot{x}_k(t) = \dot{x}_{k+1}(t) - \dot{x}_k(t)$; and $\delta = 4$ is the acceleration exponent. The reader is referred to [44] for the specifics of modeling s^* and to [45] for modifications of this model, simulation results, and links of this model with macroscopic properties of traffic flow dynamics.

1.1.2.2 Cellular Automaton (CA) Models

CA models are described by discrete equations in spatial and time coordinates. The physical space in which traffic is considered is divided into fixed-size cells, and the model is simulated in discrete increments of time. The vehicles are modeled based on certain rules including those based on "if" conditions, and at every time step of the simulation, these rules can also be updated and/or adapted to the individual vehicles. One widely studied model is the Nagel–Schreckenberg (NS) model [46], which we summarize based on [1].

The NS model starts with dividing the spatial coordinates into j number of cells, each with an equal size of Δx, and temporal coordinates into i number of equal time steps of size $\Delta t = 1$. At a time step, each cell can only be occupied by one vehicle whose speed is defined as

$$\vartheta_i = \hat{\vartheta}_i \frac{\Delta x}{\Delta t}, \tag{1.23}$$

where $\hat{\vartheta}_i \in \{0, 1, 2 \ldots, \hat{\vartheta}_{\max}\}$. After initial settings for Δx and the scaled desired velocity $\hat{\vartheta}_{\max}$, time-domain simulation is performed, where at every time step, the following rules are adapted for the vehicles:

1. *Motion*: The vehicle is moved forward by $\hat{\vartheta}_i$ number of cells.
2. *Constant acceleration*: If a vehicle is not at its maximum velocity, set $\hat{\vartheta}_i' = \hat{\vartheta}_i + 1$ while assuming constant acceleration between consecutive time steps.
3. *Constant deceleration*: If the number of cells between two consecutive vehicles is less than $\hat{d}_i$, i.e., $\hat{d}_i \leq \hat{\vartheta}_i'$, then decrease the velocity to $\hat{\vartheta}_i'' = \hat{d}_i - 1$; otherwise do not change the velocity, $\hat{\vartheta}_i'' = \hat{\vartheta}_i'$.
4. *Randomization*: With a probability p, whenever $\hat{\vartheta}_i'' \geqslant 1$, reduce the velocity to $\hat{\vartheta}_{i+1} = \hat{\vartheta}_i'' - 1$; otherwise, set $\hat{\vartheta}_{i+1} = 0$.

In light of the last three steps, one can formulate the updated velocity of each vehicle at any time step as

$$\hat{\vartheta}_{i+1} = \max(0, \min(\hat{\vartheta}_{\max}, \hat{d}_i - 1, \hat{\vartheta}_i + 1) - \xi_i^{(p)}), \tag{1.24}$$

where $\xi_i^{(p)} = 1$ is a Bernoulli random variable, which is 1 with probability p and 0 with probability $1 - p$. When studying freeway traffic, it is suggested that $p = 0.5$, $\Delta x = 7.5\,\text{m}$, and $\hat{\vartheta}_{\max} = \vartheta_0 \Delta t / \Delta x = 5$. For city traffic, the suggested parameters are $p = 0.2$ and $\hat{\vartheta}_{\max} = 2$.

Further remarks: CA models could be classified separately from microscopic car-following models as CA models are discrete-time "particle hopping" models, while car-following models are based on continuous-time nonlinear differential equations. Nevertheless, in both modeling efforts, the behavior of every single vehicle (or, particle) is updated over time and space based on a particular update law; hence in this sense, we have presented both modeling approaches under the "microscopic model" category.

CA models, which are also called coupled-map lattice models, are often preferred over continuous models when faster computation is critical, especially when performing simulations with many vehicles. This advantage is appealing for the implementation of such models in real-life traffic simulations [47]. CA models are also attractive when one would like to incorporate some degree of stochasticity and/or if-conditions in the model. On the other hand, a possible challenge with CA models is that it may be difficult to utilize them based on a mathematical theory with the intention to grasp analytical results and to capture underlying rules of traffic flow. CA models may also not be preferred in cases when one wishes to model vehicle dynamics and human driver behavior with a certain degree of detail. In such cases, it may be preferable to utilize microscopic models [1, p. 1088].

Besides studying highway traffic, CA models can also be utilized to optimize traffic lights in urban traffic. In [48], a representation of a North–South and East–West bounds of a city traffic is studied under different traffic-light-signalization patterns to investigate traffic synchronization.

While our presentation briefly summarized the NS model, the topic of CA modeling is rich. For example, the Biham–Middleton–Levine (BML) model [49] uncovers jamming transitions from low- to high-density traffic flow and explains how vehicles self-organize in these two different states of traffic [49]. See also the surveys [1, 2] for additional discussions on CA models.

1.1.3 Gas-Kinetic Models

Gas-kinetic models can provide a foundation to link microscopic models with macroscopic ones [1]. In this approach, the kinetic theory of gases are adopted to study traffic flow assuming the vehicles in traffic act similarly to interacting particles in a gas flow [2].

For the traffic problem, the gas kinetics equations need to be adjusted appropriately. In the study of gas kinetics, particles share their momentum by colliding with each other, while in traffic, collisions have to be strictly avoided. Some earlier models based on kinetic theories are attributed to Prigogine [17]. Defining $f(x, \vartheta, t)dx\, d\vartheta$

as the number of vehicles at time t between positions x and $x + dx$, and between velocities ϑ and $\vartheta + d\vartheta$, one can express the continuity equation analogously to the Boltzmann equation in kinetic theory of gases,

$$\frac{\partial f}{\partial t} + \vartheta \frac{\partial f}{\partial x} = \left(\frac{\partial f}{\partial t}\right)_{\text{rel}} + \left(\frac{\partial f}{\partial t}\right)_{\text{int}}, \tag{1.25}$$

where the second term on the right-hand side corresponds to the interactions between vehicles, and the first term $(\partial f/\partial t)_{\text{rel}}$ accounts for the relaxation of f to f_{des} in the absence of mutual interactions, given by

$$\left(\frac{\partial f}{\partial t}\right)_{\text{rel}} = -\frac{f - f_{\text{des}}}{\alpha_{\text{rel}}}, \tag{1.26}$$

where α_{rel} is the relaxation time. In addition, f_{des} is considered as

$$f_{\text{des}}(x, \vartheta, t) = c(x, t) F_{\text{des}}(\vartheta), \tag{1.27}$$

where $F_{\text{des}}(\vartheta)$, the desired speed distribution, is decoupled from the local density $c(x, t)$. Combining (1.25), (1.26), and (1.27), the Boltzmann-like Prigogine equation is obtained

$$\frac{\partial f}{\partial t} + \vartheta \frac{\partial f}{\partial x} = -\frac{f(x, \vartheta, t) - c(x, t) F_{\text{des}}(\vartheta)}{\alpha_{\text{rel}}} + \left(\frac{\partial f}{\partial t}\right)_{\text{int}}. \tag{1.28}$$

The above model is valid for low densities and can be further extended for higher densities [1, Sect. III.E]. The model is systematically studied by Paveri-Fontana [50] specifically from the viewpoint of how $(\partial f/\partial t)_{\text{rel}}$ is formulated, and several improvements are presented over Prigogine's model.

Further remarks: As discussed in [2], most kinetic models are concerned with single-lane traffic flow, yet extensions to multi-lane traffic are possible based on formulating gas-kinetic models for each lane and modeling the characteristics of lane changing. For example, in [51], the authors propose that vehicles in each lane obey a Boltzmann-like interaction term while lane-changing dynamics are modeled by overtaking probabilities and waiting times of lane-change maneuvers. Gas-kinetic models can also reflect the effects of different types of vehicles. In [52], the authors demonstrate a technique for taking into account the heterogeneous nature of vehicles. They also show how to convert gas-kinetic equations to macroscopic models for computational efficiency of time simulations. Indeed, one can bridge gas-kinetic models to microscopic and macroscopic scales. In [53], a thermodynamic approach is taken where headway and inverse of time-delay $1/\tau$ of the time-delayed Optimal Velocity Model[6] are characterized as their counterparts, respectively, the order parameter and temperature, based on the time-dependent Ginzburg–Landau (TDGL) equation.

[6] Microscopic models with time-delays are covered in the next section; see Eq. (2.7).

The results allow to create a phase diagram on the plane of headway versus $1/\tau$ to explain jamming transitions, and moreover, the connection between TDGL equation and Korteweg–de Vries (KdV) equation is established. The study also presents a linear stability analysis as well as validation of phase diagrams by numerical simulations. For bridging microscopic and macroscopic models, we also cite [11], where authors take Bando's Optimal Velocity Model (OVM) (1.16) with a particular optimal velocity function V, and derive a corresponding macroscopic model of OVM. Along these lines, in [54], the authors bridge macroscopic parameters with the model in (1.15) including a time delay, and provide comparisons with experimental data.

1.2 Experiments and Software

In parallel to modeling and analytical work, many experimental studies have also been reported. For example, Edie investigated experimental data and showed that *Density-Flow* graphs are formed by two separate branches, where one branch corresponds to free flow and the other to congested flow [55]. It was reported that these two branches are separated by a discontinuity; specifically that flow will increase up to a critical density, but the flow, or capacity, will drop when the density further grows in traffic. The study in [56] reveals from experimental data the characteristics of stop-and-go traffic flow and mechanisms of self-organization of vehicles in traffic. There the author presents a typical flow-density diagram, which resembles the mirror image of a reversed "λ symbol" [56, Fig. 2a]. This pattern is also reported in [57] where the authors present a comparison between the OVM in (1.16) and experimental data, showing that the OVM satisfactorily produces the capacity drop discontinuity on the flow-density graph.

In [45], adaptation of driving style and memory effect are incorporated into the driver modeling and implemented in the Intelligent Driver Model (IDM) (1.22); the resulting simulations of traffic flow are shown to be in good agreement with empirical data. Other studies of the authors include an in-depth study, comparing experimental results and numerical simulations [44]. In time simulations, authors utilize the IDM (1.22) combined with empirical boundary conditions and physical obstacles, such as heterogeneities, and present validation results between models and empirical data.

While the studies in [44, 45, 54, 56, 57] are in principle inspired by microscopic models, they also draw conclusions regarding the macroscopic features of measured traffic flow data. Validation between models and data is performed on flow-density graphs; see also [58]. Some of the work on experimental data is validated directly on macroscopic characteristics. Experimental results and their consequences on the macroscopic properties are given in [59]. Similarly, experimental results have been used in [60, 61] to identify and characterize the phase transitions pertaining to various behavioral characteristics of traffic flow. In [61], the author investigates the synchronized flow patterns versus general flow patterns in traffic, and *weak* versus *strong congestion* properties of traffic dynamics, while also establishing the link between *wide moving jams* and *synchronized flow* phenomenon.

Measurement of data can be performed by sensors located at fixed points on highways and in the streets, where data pertains to, for example, the status of traffic light switches, speed of vehicles, and spacing between vehicles. Data about traffic can be utilized to improve energy efficiency [62] and, via data-driven tools, can also be used to understand the complex traffic flow dynamics and perform robust forecasting [63]. With the emergence of autonomous vehicles, data pertaining to the states of these vehicles are also available and can be utilized to improve traffic flow metrics [64]. Studies also exist where controlled experiments are performed with instrumented vehicles [65] and small-scale robotic vehicles [66].

1.2.1 *Software Development*

Since analytical or explicit solutions may not be available in certain cases, computer simulations are also necessary. Various freeware and commercial simulation tools are developed and used for this purpose, which are aimed at simulating traffic flow dynamics based on models of macroscopic, microscopic, or gas-kinetic type.

For instance, a microscopic model known as *non-integer car-following model* (see [1, Sect. III.A.1] and Chap. 2.2 below) is used in the traffic flow simulation package MITSIM [67]. Payne's model [68], as discussed in [1], combines the LW model with an optimal velocity model in which, as proposed in [69], the optimal velocity function is affected by a time delay. This model was utilized to develop the freeway macroscopic simulation package FREFLO. A simulation package based on a *nonlocal, gas-kinetic-based traffic model* is reported in [70]. See also the TRANSYT [71], SCOOT [72], and CRONOS [73] packages. For further examples, e.g., from cellular automaton models, see the simulation package TRANSIMS [74] and the web page of Dr. Martin Treiber [75], which presents 3D graphic simulations and multi-lane traffic behavior. Designed and developed as a Java applet by Treiber, a type of car-following model (1.22) is utilized to predict traffic flow behavior including various parameters, such as the presence of lane closures, multiple lanes, and on-ramp. On fuel economy, we cite the simulator AUTONOMIE [76].

1.3 Notes and Further Reading

We briefly presented several ideas behind mathematical models of traffic flow dynamics. The topic is indeed broad, and it is not possible to capture every detail within a brief; yet we hope this chapter demonstrates approaches to modeling with the references provided. Discussions on macroscopic models, cellular automaton, and gas-kinetic models are presented with their general characteristics only, since excellent surveys and books exist on these topics [1, 2, 20, 21]. Readers will also find the work in [77] insightful on the most recent modeling results and integration of publicly available data. Moreover, we aim to provide a historical perspective to modeling while acknowledging the richness of results in the literature, which impact even today's research in transportation engineering. The main goal of this brief is to understand, through stability theory, how time-delays in microscopic models affect various traffic flow dynamics and how to analyze such effects on the qualitative properties of the models' dynamics. To this end, in the remainder of this brief, we focus on *microscopic models, time-delays*, and *stability*. Some content in this chapter was reprinted/adapted by permission from Springer Nature: Springer-Verlag, from the chapter *Mathematical models of traffic flow dynamics with time delays* by R. Sipahi and S.-I. Niculescu, published in *Complex Time-Delay Systems: theory and applications* (ed. Fatihcan M. Atay), Copyright 2009 [78]. The presentation here was updated with various references and discussions and completed with insights.

References

1. D. Helbing, Traffic and related self-driven many-particle systems. Rev. Mod. Phys. **73**, 1067–1141 (2001)
2. D. Chowdhury, L. Santen, A. Schadschneider, Statistical physics of vehicular traffic and some related systems. Phys. Rep. **329**, 199–329 (2000)
3. H.Y. Lee, H.-W. Lee, D. Kim, Origin of synchronized traffic flow on highways and its dynamic phase transitions. Phys. Rev. Lett. **81**, 1130–1133 (1998)
4. H.Y. Lee, H.-W. Lee, D. Kim, Dynamic states of a continuum traffic equation with on-ramp. Phys. Rev. E **59**, 5101–5111 (1999)
5. M.J. Lighthill, G.B. Whitham, On kinematic waves. II. a traffic flow theory on long crowded roads. Proc. Roy. Soc. Lond. Ser. A **229**, 317–345 (1955)
6. B.D. Greenshields, J. Rowland Bibbins, W.S. Channing, H.H. Miller, A Study of Traffic Capacity. Highway Research Board, vol. 14 (1935) pp. 448–477
7. J.M. Burgers, *The Nonlinear Diffusion Equation: Asymptotic Solutions and Statistical Problems* (Springer Dordrecht, 1974)
8. D. Bilodeau, Cole-Hopf Transformation. http://www.math.mcgill.ca/gantumur/math580f11/downloads/ColeHopf.pdf. Accessed 9 June 2023
9. G.B. Whitham, *Lectures on Wave Propagation* (Springer, Berlin, 1979)
10. E. Hopf, The partial differential equation $u_t + uu_x = \mu_{xx}$. Commun. Pure Appl. Math. **3**, 201–230 (1950)
11. P. Berg, A. Mason, A. Woods, Continuum approach to car-following models. Phys. Rev. E **61**, 1056–1066 (2000)
12. T. Nagatani, Jamming transition in a two-dimensional traffic flow model. Phys. Rev. E **59**, 4857–4864 (1999)
13. D. Helbing, From microscopic to macroscopic traffic models, in *A Perspective Look at Nonlinear Media. From Physics to Biology and Social Sciences*, ed. by J. Parisi, S.C. Muller, W. Zimmermann, vol. 503 (Springer, Berlin, 1998)
14. P. Nelson, Synchronized traffic flow from a modified Lighthill-Whitman model. Phys. Rev. E **61**, R6052–6055 (2000)
15. D. Helbing, B. Tilch, Generalized force model of traffic dynamics. Phys. Rev. E **58**, 133–138 (1998)
16. W.F. Phillips, A new continuum traffic model obtained from kinetic theory, in *1978 IEEE Conference on Decision and Control, SanDiego, CA* (1979), pp. 1032–1036

17. I. Prigogine, F.C. Andrews, A Boltzmann-like approach for traffic flow. Oper. Res. **8**(6), 789–797 (1960)
18. B.S. Kerner, S.L. Klenov, P. Konhäuser, Asymptotic theory of traffic jams. Phys. Rev. E **56**, 4200–4126 (1997)
19. G. Orosz, Connected cruise control: modelling, delay effects, and nonlinear behaviour. Veh. Syst. Dyn. **54**(8), 1147–1176 (2016)
20. M. Treiber, A. Kesting, *Traffic Flow Dynamics: Data, Models and Simulation* (Springer, Berlin, Heidelberg, 2013)
21. B.S. Kerner, *Introduction to Modern Traffic Flow Theory and Control: The Long Road to Three-Phase Traffic Theory* (Springer, Berlin, 2009)
22. A. Bose, P.A. Ioannou, Analysis of traffic flow with mixed manual and semi automated vehicles. IEEE Trans. Intell. Transp. Syst. **4**(4), 173–188 (2003)
23. L.A. Pipes, An operational analysis of traffic dynamics. J. Appl. Phys. **24**(3), 274–281 (1953)
24. M. Bando, K. Hasebe, A. Nakayama, A. Shibata, Y. Sugiyama, Dynamical model of traffic congestion and numerical simulation. Phys. Rev. E **51**(2), 1035–1042 (1995)
25. M. Bando, K. Hasebe, K. Nakanishi, A. Nakayama, Analysis of optimal velocity model with explicit delay. Phys. Rev. E **58**, 5429–5435 (1998)
26. K. Konishi, H. Kokame, K. Hirata, Coupled map car-following model and its delayed-feedback control. Phys. Rev. E **60**, 4000–4007 (1999)
27. S. Kurata, T. Nagatani, Enhancement and stabilization of traffic flow by moving in groups. Phys. Rev. E **64**, 016106 (2001)
28. I. Gasser, G. Sirito, B. Werner, Bifurcation analysis of a class of 'car following' traffic models. Phys. D **197**, 222–241 (2004)
29. A.D. Mason, A.W. Woods, Car-following model of multispecies systems of road traffic. Phys. Rev. E **55**, 2203–2214 (1997)
30. H.J.C. Huijberts, Improved stability bound for steady-state flow in a car-following model of road traffic on a circular route. Phys. Rev. E **65**, 047103 (2002)
31. V.I. Shvetsov, Mathematical modeling of traffic flows. Autom. Remote. Control. **64**, 1651–1689 (2003)
32. K. Nakanishi, K. Itoh, Y. Igarashi, M. Bando, Solvable optimal velocity models and asymptotic trajectory. Phys. Rev. E **55**, 6519–6532 (1997)
33. T. Nagatani, Stabilization and enhancement of traffic flow by the next-nearest-neighbor interaction. Phys. Rev. E **60**, 6395–6401 (1999)
34. H. Lenz, C.K. Wagner, R. Sollacher, Multi-anticipative car-following model. Eur. Phys. J. B **7**, 331–335 (1999)
35. A. Nakayama, Y. Sugiyama, K. Hasebe, Effect of looking at the car that follows in an optimal velocity model of traffic flow. Phys. Rev. E **65**, 016112 (2001)
36. H.X. Ge, S.Q. Dai, L.Y. Dong, Y. Xue, Stabilization effect of traffic flow in an extended car-following model based on an intelligent transportation system application. Phys. Rev. E **70**, 066134 (2004)
37. T. Nagatani, Traffic jams induced by fluctuation of a leading car. Phys. Rev. E **61**, 3534–3540 (2000)
38. R. Nagai, T. Nagatani, A. Yamada, Phase diagram in multi-phase traffic model. Phys. A **355**, 530–550 (2005)
39. R. Nagai, T. Onouchi, T. Nagatani, Phase separation and evolution of one pulse jam in traffic flow. Phys. A **354**, 571–581 (2005)
40. T. Bleile, A new microscopic model for car-following behaviour in urban traffic, in *Mobility for Everyone. 4th World Congress on Intelligent Transport Systems, Oct 21–24 1997, Berlin*. (Paper No. 2056) (1997)
41. D. Helbing, *Quantitative Sociodynamics* (Springer, Dordrecht, 1995)
42. D. Helbing, P. Molnar, Social force model for pedestrian dynamics. Phys. Rev. E **51**, 4282–4286 (1995)
43. R. Jiang, Q. Wu, Z. Zhu, Full velocity difference model for a car-following theory. Phys. Rev. E **64**, 017101 (2001)

44. M. Treiber, A. Hennecke, D. Helbing, Congested traffic states in empirical observations and microscopic simulations. Phys. Rev. E **62**, 1805–1824 (2000)
45. M. Treiber, D. Helbing, Memory effects in microscopic traffic models and wide scattering in flow-density data. Phys. Rev. E **68**, 046119 (2003)
46. K. Nagel, M. Schreckenberg, A cellular automaton model for freeway traffic. J. Phys. I **2**, 2221–2229 (1992)
47. K. Nagel, Particle hopping models and traffic flow theory. Phys. Rev. E **53**, 4655–4672 (1996)
48. E. Brockfeld, R. Barlovic, A. Schadschneider, M. Schreckenberg, Optimizing traffic lights in a cellular automaton model for city traffic. Phys. Rev. E **64**, 056132 (2001)
49. O. Biham, A.A. Middleton, D. Levine, Self-organization and a dynamical transition in traffic-flow models. Phys. Rev. A **46**, R6124 (1992)
50. S.L. Paveri-Fontana, On Boltzmann-like treatments for traffic flow: a critical review of the basic model and an alternative proposal for dilute traffic analysis. Transp. Res. **9**, 225–235 (1975)
51. D. Helbing, A. Greiner, Modeling and simulation of multilane traffic flow. Phys. Rev. E **55**, 5498–5508 (1997)
52. D. Helbing, Modeling multi-lane traffic flow with queuing effects. Phys. A **242**, 175–194 (1997)
53. T. Nagatani, Thermodynamic theory for the jamming transition in traffic flow. Phys. Rev. E **58**, 4271–4276 (1998)
54. D.C. Gazis, R. Herman, R.B. Potts, Car-following theory of steady-state traffic flow. Oper. Res. **7**(4), 499–505 (1959)
55. L.C. Edie, Car-Following and steady-state theory for noncongested traffic. Oper. Res. **9**, 66–76 (1961)
56. B.S. Kerner, Experimental features of self-organization in traffic flow. Phys. Rev. Lett. **81**, 3797–3800 (1998)
57. M. Bando, K. Hasebe, K. Nakanishi, A. Nakayama, A. Shibata, Y. Sugiyama, Phenomenological study of dynamical model of traffic flow. J. Phys. I. **5**, 1389–1399 (1995)
58. B.S. Kerner, S.L. Klenov, Microscopic theory of spatial-temporal congested traffic patterns at highway bottlenecks. Phys. Rev. E **68**, 036130 (2003)
59. B.S. Kerner, H. Rehborn, Experimental properties of complexity in traffic flow. Phys. Rev. E **53**, R4275 (1996)
60. B.S. Kerner, H. Rehborn, Experimental properties of phase transitions in traffic flow. Phys. Rev. Lett. **79**, 4030–4033 (1997)
61. B.S. Kerner, Empirical macroscopic features of spatial-temporal traffic patterns at highway bottlenecks. Phys. Rev. E **65**, 046138 (2002)
62. G. Mahler, A. Vahidi, An optimal velocity-planning scheme for vehicle energy efficiency through probabilistic prediction of traffic-signal timing. IEEE Trans. Intell. Transp. Syst. **15**, 2516–2523 (2014)
63. A.M. Avila, I. Mezić, Data-driven analysis and forecasting of highway traffic dynamics. Nat. Commun. **11**(1), 2090 (2020)
64. T. Ersal, I. Kolmanovsky, N. Masoud, N. Ozay, J. Scruggs, R. Vasudevan, G. Orosz, Connected and automated road vehicles: state of the art and future challenges. Veh. Syst. Dyn. **58**(5), 672–704 (2020)
65. I.G. Jin, S.S. Avedisov, C.R. He, W.B. Qin, M. Sadeghpour, G. Orosz, Experimental validation of connected automated vehicle design among human-driven vehicles. Transp. Res. Part C: Emerg. Technol. **91**, 335–352 (2018)
66. W.B. Qin, G. Orosz, Experimental validation of string stability for connected vehicles subject to information delay. IEEE Trans. Control Syst. Technol. **28**(4), 1203–1217 (2019)
67. Q.I. Yang, H.N. Koutsopoulos, A microscopic traffic simulator for evaluation of dynamic traffic management systems. Transp. Res. Part C: Emerg. Technol. **4**(3), 113–129 (1996)
68. H.J. Payne, FREFLO: a macroscopic simulation model of freeway traffic. Transp. Res. Rec. **722** (1979)
69. G.F. Newell, Nonlinear effects in the dynamics of car following. Oper. Res., **9**, 209–229 (1961)
70. D. Helbing, M. Treiber, Numerical simulation of macroscopic traffic equations. Comput. Sci. Eng. **1**, 89–98 (1999)

71. TRANSYT. https://trlsoftware.com/products/junction-signal-design/transyt/. Accessed 9 June 2023
72. SCOOT. https://trlsoftware.com/products/traffic-control/scoot/. Accessed 9 June 2023
73. F. Boillot, J.M. Blosseville, J.B. Lesort, V. Motyka, M. Papageorgiou, S. Sellam, Optimal signal control of urban traffic networks, *Road Traffic Monitoring, (IEEE Conference Publication 355)* (IET, 1992), p. 75–79
74. TRANSIM. https://www.fhwa.dot.gov/planning/tmip/resources/transims/index.cfm. Accessed 9 June 2023
75. Simulators by Dr. Martin Treiber. http://www.mtreiber.de/movie3d/index.html, https://www.traffic-simulation.de/. Accessed 9 June 2023
76. https://www.autonomie.net. Accessed 9 June 2023
77. L. Li, R. Jiang, Z. He, X.M. Chen, X. Zhou, Trajectory data-based traffic flow studies: a revisit. Transp. Res. Part C **114**, 225–240 (2020)
78. R. Sipahi, S.-I. Niculescu, Mathematical models of traffic flow dynamics with time delays, in *Complex Time-Delay Systems: theory and applications*, ed. by Fatihcan M. Atay (Springer, 2010)

Chapter 2
Delays in Traffic Flow Models

In this chapter, we present how time-delays arise in traffic flow dynamics, mainly focusing on *human reaction delays*, *mechanical delays*, and *delay time of vehicle motion*. We clarify the differences in notions of time constant versus time-delay, also known as dead-time, and provide definitions of discrete delays and distributed delays. This is followed by a quantitative measure of time-delay as reported in the literature and as measured under certain driving conditions. We give examples of linear and nonlinear microscopic traffic flow models that include time-delays, with various discussions, comparisons, and interpretations. Additional insights on driver behavior modeling can be found in [1].

2.1 Time-Delay Effects in Traffic Flow Dynamics

In order to study time-delay effects on traffic flow dynamics, it is important to understand how delays arise in the context of traffic flow dynamics, what their magnitudes might be under various conditions, and how their presence could affect mathematical models. Since delays arise mainly from human reactions and vehicle mechanical response, it is natural to model such effects by considering individual drivers and vehicles, which calls for microscopic models. We furthermore deal mainly with continuous states and deterministic models in order to provide a consistent presentation. We refer the interested reader to CA models for studying stochastic effects, and to existing studies on stochastic systems mentioned in Chap. 1.

2.1.1 *Origin of Delays*

Time-delays in the context of traffic dynamics are classified first according to their origin. In general, time-delay τ represents a time-varying physical entity, i.e., $\tau =$

R. Sipahi et al., *Deterministic Car-Following Traffic Models*,
SpringerBriefs in Control, Automation and Robotics,
https://doi.org/10.1007/978-3-031-58164-9_2

$\tau(t)$. Moreover, delays can be quantitatively different for each driver, and can be considered to be the sum of two terms, a *time-invariant part*, which is constant and denoted by τ_{inv}, and a *time-varying part* denoted by $\tau_v(t)$. Based on the discussions and characterizations in [2–5], time-delay $\tau(t)$ in traffic dynamics can be considered in three parts,[1] as outlined below.

2.1.1.1 Physiological Time Lag

Physiological time lag arises due to human drivers, hence the attribute *physiological*. When the driver k receives a stimulus from the preceding vehicle $k+1$ in the form of cognitive information about the variations of the headway $\Delta x_k(t) = x_{k+1}(t) - x_k(t)$ and velocities $\dot{x}_{k+1}(t) - \dot{x}_k(t)$, they perform an action on vehicle k in the way of acceleration (or deceleration) by applying the gas pedal (or the brake pedal). However, a human's reaction to such stimulus does not occur instantaneously, but requires a finite period of time $\bar{\tau}$, which is called the *physiological lag*. One can further decompose the physiological delay into a time-invariant $\bar{\tau}_{inv}$ and a time-varying part $\bar{\tau}_v(t)$. That is, $\bar{\tau}(t) = \bar{\tau}_{inv} + \bar{\tau}_v(t)$.

Human reaction delays, or delays in human physiology in general, have been extensively studied in human factors engineering, and such delays are of importance in any human–machine interaction system including vehicle driving. Earlier studies in this direction are rich; for example, [6] includes a comprehensive compilation of literature, summarizing the results of human perception, decision-making, and the influence of the environment on how humans decide, and how quickly and accurately they decide. It also reports human reaction times against visual versus auditory signals and the outcomes of human decision-making when faced with multiple choices [6]. Along these lines, human workload modeling in connection with human sensory system and memory has been studied in [7, 8], focusing on information processing and its relation to human performance in executing a task on a machine. Closely related to the discussion below, we cite [4, 9–12] in the context of human drivers, which present various human perception theories. The Ref. [4] includes a comprehensive list of human reaction times under various experimental conditions, including those compiled from the literature.

In the sequel, we base our discussions primarily from the contributions in [4]. Accordingly, physiological lags are categorized into two main parts: the so-called (a) *mental processing time* and (b) *movement time*. Mental processing time is further broken down into three subcategories:

[1] We caution the reader that some definitions of time-delays in the literature vary and may not always be consistent. A discussion can be found in [4, p. 200].

(a1) *Sensing time*. This is the time it takes for a human driver to *detect an object*, to realize that "there is an object". Under fixed conditions, it is known that human reaction time decreases when the object is presented with stronger signal intensity, which, for example, includes brightness, size, and visibility. It turns out that human reaction times are *smaller* for auditory signals than they are for visual ones, and if auditory and visual signals are combined together with proper timing, human reaction times can be decreased [6, p. 1152].

(a2) *Perception time*. An additional amount of time is needed to perceive *what* the object is. For instance, after the sensing time elapses, the driver would say that "the shape in the street is a bicycle". Perception time increases under uncertainty, low visibility, and surprise. Moreover, when faced with multiple choices against a stimulus, reaction time due to perception can increase. See [6, p. 1154] where simple or choice reaction time results are presented.

(a3) *Response selection and programming time*. This is the time needed for the driver to "mentally program the movement" after the object has been seen and perceived. For instance, the driver could decide to steer the vehicle instead of braking. Response selection also requires more time when the driver has to evaluate multiple possible responses. For example, driving in a city traffic where many parameters affect driving, such as pedestrians, traffic lights, bicycles, and lane changes, it is expected that response selection and programming will take longer than it would on highways. With practice, this delay time can be decreased.

(b) *Movement time*. This is the amount of time it takes the *driver's neuromuscular system to perform the programmed movement*, that is, to produce a physical movement of arms, legs, and head, for operating the vehicle. An example is the time it takes to move the foot from one pedal to another. Movement time is generally larger for complex movements, but can be decreased with increased focus level and practice.

2.1.1.2 Mechanical Time Lag

This delay time τ_m is the time that elapses between when the driver applies the gas/brake pedal and when the vehicle starts to accelerate/decelerate. It is determined by the mechanical capabilities of the vehicle and is independent of the human operator. Unlike physiological delays, this delay may be considered time-invariant for a vehicle, though it may vary across vehicles. Mechanical delay occurs sequentially followed by the physiological delay in an acceleration/deceleration action.[2]

[2] In some studies physiological and mechanical time lags are lumped together as a single time-delay τ; see, for instance, [2, 3], where τ is denoted as the *delay time of response*, and the models surveyed in Sect. 2.2.

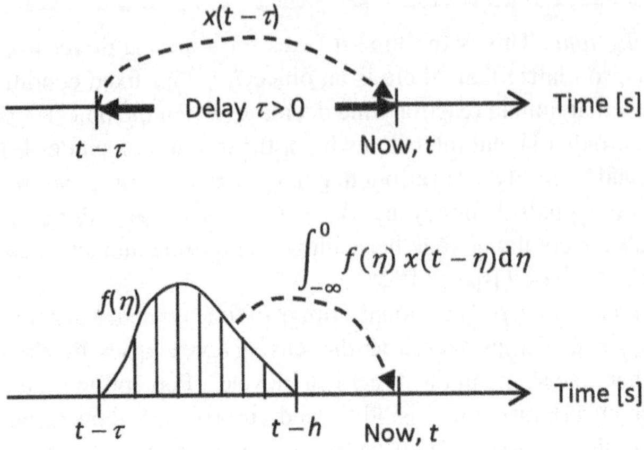

Fig. 2.1 Discrete delay (top) versus distributed delay (bottom). In the case of a discrete delay, the information from a single time point $t-\tau$ in the past affects the dynamics at the present time t. In the case of a distributed delay, a collection of such effects over the whole time interval $[t-\tau, t-h]$ in the past collectively affects the present dynamics. These past values may have different weights, as specified by a distribution function $f(\eta)$. See [14] for technical discussions

2.1.1.3 Delay Time of Vehicle Motion

Another notion of delay is the delay time of vehicle motion T; see, e.g., [3], which is the time a follower vehicle needs to bring its speed to the same velocity of its leader. This time is inherent to the dynamics of the vehicles. In a linear systems setting, T corresponds to the period of time between a sinusoidal excitation input $M \sin(\omega t)$ and its sinusoidal response $\bar{M} \sin(\omega t - \phi)$ in steady state. In other words, $T = \phi/\omega$. In this setting, $\bar{M}$ and ϕ are determined directly from the corresponding vehicle model, which in turn determines T as a function of the excitation frequency ω. As discussed in [3], T is different from physiological and mechanical time lags since it is not a *pure delay* or *dead-time* (see also Fig. 2.1). Hence, even in the absence of physiological and mechanical time lags, T exists. Nevertheless, in the presence of such lags, the vehicle model changes and thus T is also affected by such lags.

2.1.2 *Discrete Delays and Distributed Delays*

A formal mathematical introduction of delays and delay differential equations is provided in Chap. 3. Here we define two types of delays, namely *discrete* and *distributed delays*, in order to use them in our discussion of the origin of delays. This discussion will help us understand how the presence of delays could modify the microscopic car following models and prepare the basis for the following chapters.

In principle, due to human reaction delays, an action performed by a driver is based on what was experienced a certain time ago in the form of a stimulus. This is the delay time $\tau > 0$. In this setting, for instance, when a stimulus in the form of headway $x_{k+1}(t) - x_k(t)$ is sensed by the driver k at time t, the driving decision based on this stimulus will be produced in the future at time $t + \tau$. Equivalently, the decision made at time t depends on an event at time $t - \tau$. For this reason, this type of delay is known as *discrete delay*, which is a time shift operator, sometimes called a *pure delay* or *dead-time* (Fig. 2.1). In light of the above discussions, we can consider the *physiological lag* and *mechanical time lag* as *discrete delays*, while the delay time of vehicle motion is related to the bandwidth of the vehicle.[3] (See Sect. 2.1.1 for definitions.)

In some cases, the stimulus arrives from a collection of time points in the past instead of a single point. In the limit, such a collection would contain infinitely many points distributed over an interval in the past, all contributing to shaping a decision made at time t. Two examples of this are found in modeling driver's memory effects and in the design of automated controllers. When a driver is traveling in traffic, they continuously experience the traffic scene and form decisions based on this continuous experience. In other words, the driver's decisions are influenced by a collection of experiences collected from a continuum of time points in the past, but limited by the memory horizon of the driver. In the case of automated controllers, one could argue that more information would lead to better decisions, and hence, if delays cannot be avoided, it may be better to collect as much past information as possible and use it in making decisions. Motivated by this rationale, one can design an automated controller for headway spacing control between consecutive vehicles based on headway information collected over a time interval in the past. See also [13] for further motivations and the foundation of distributed delays.

To give an example, one can incorporate a discrete delay into the model (1.15) as

$$\dot{v}_k(t) = \alpha_k(v_{k+1}(t - \tau) - v_k(t - \tau)), \tag{2.1}$$

which describes the accelerations of vehicle k based on the stimulus of speed variations as sensed by the driver τ seconds ago, and acted upon proportionally by a strength α_k. More generally, the model (2.1) can be converted to capture memory effects in a system of equations [15],

$$\dot{v}_k(t) = \alpha_k \int_0^\infty f(\tau)(v_{k+1}(t - \tau) - v_k(t - \tau))\, d\tau, \tag{2.2}$$

where f is the *delay kernel* (see Fig. 2.1 and Chap. 5)—a measurable function of exponential order, which represents the relative importance of the past data points of velocity differences. For example, when the kernel is a *uniform distribution* over an interval, all the past information from that time interval have equal importance

---

[3] In certain instances τ is referred to as the human reaction delay; however, one can also consider τ as the lump sum of all sequential time lags, including the mechanical time lag.

in terms of creating a decision on the right-hand side of (2.2). Leaving the details to Chap. 3, we close this part by noting that when f is a Dirac delta function, (2.2) reduces to the discrete delay model (2.1).

2.1.3 Quantitative Measure of Time-Delay

In [16], the author benchmarks several car-following models to investigate model sensitivity against collision. Specifically, a platoon of up to 200 vehicles is considered with different human reaction times in the range 0.5–2.0 s (see Sect. 2.2 for further details on the models). Delay values utilized in [2] are in the range 0.1–0.4 s. In [4], the braking response time of drivers against common but uncertain signals is given in the range 1.2–1.35 s with a standard deviation of 0.6 s. Such delays are shorter for expected signals, in the range of 0.70–0.75 s, but they double to 1.5 s in the event of a surprise. Further details on delay magnitude can be found in Green's work [4, pp. 197–199], with a detailed table covering a variety of different conditions under which reaction delay measurements were performed with human subjects. The cited study reports that reaction times are in the range of 0.6–1.9 s depending on a multitude of factors. Moreover, the American Association of State Highway and Transportation Officials suggests a 2.5 s braking response time for drivers under normal road conditions and visibility [4, p. 208], which is a consistent and safe response time, given the above measures of delays.

The delay times in longitudinal and lateral control of vehicles can also vary. In [17, Fig. 2], a comparison is given between longitudinal speed control associated with the desire to remain in the lane and the steering control associated with the desire to change lanes. Overall, steering control is associated with an order of magnitude higher bandwidth than longitudinal speed control [17]. Nevertheless, it is not always safe to perform steering control especially in crowded traffic and at higher cruising speeds. Experimental results collected in driving simulations are analyzed based on the Fourier transform, which indicate human reaction times in the range of 1–2 s. An interesting comparison is provided in the study regarding sensing times, where a sensing time of 1.9 s is needed if the preceding vehicle accelerates at 0.76 m/s^2, while sensing time increases to 2.5 s for sensing an acceleration of 0.5 m/s^2 of the preceding vehicle. As a general remark, it is stated that the response time of a human operator decreases by about 0.8 s for an increase in acceleration by 0.3 m/s^2 [17, p. 168].

2.2 Examples of Time-Delayed Microscopic Traffic Flow Models

We provide several microscopic traffic flow models in which delays are explicitly considered. The list is surely incomplete, but it includes various well-known models and guides the reader through developing models with certain characteristics. This chapter should also be viewed in connection with Chap. 1 where models without delays are discussed. For causality reasons, we adapt all models from the literature to the case where position and velocities arriving from the past $t-\tau$ and influence accelerations at the current time t.

2.2.1 A Model by Chandler

Chandler's model seems to be the first model with a time-delay term published in the literature [18]. It can be considered as the time-delayed version of Reuschel's delay-free model published in 1950 [19, 20] (see also Eq. (1.15)) and has the form

$$\ddot{x}_k(t) = \alpha(\dot{x}_{k+1}(t-\tau) - \dot{x}_k(t-\tau)). \tag{2.3}$$

Here α is the sensitivity of the driver as defined for (1.15), and τ is the *constant human reaction delay*, representing the time needed by the drivers to accelerate or decelerate their vehicles based on the velocity differences sensed τ sec ago. Notice in the above model that the order can be reduced by one, with a simple change of variables as $v_k = \dot{x}_k$, where v_k represents the velocity of vehicle k.

Chandler's model offers a good balance of complexity and simplicity, and consequently has been widely used in various studies. For instance, a 2004 report by the University of Southern California [21] incorporates a constant delay time into the delay-free model of Pipe [22], which indeed recovers Chandler's model [18], and simulates a mixed traffic of trucks and cars.

2.2.2 A Model by Gazis

In [23], the following nonlinear model is introduced:

$$\ddot{x}_k(t) = \frac{\alpha}{x_{k+1}(t-\tau) - x_k(t-\tau)}(\dot{x}_{k+1}(t-\tau) - \dot{x}_k(t-\tau)), \tag{2.4}$$

where the added denominator creates a headway sensitive driver, who has τ seconds of reaction time. That is, the driver makes more aggressive acceleration/deceleration decisions when the headway decreases. This model captures the fact that a driver indeed becomes more attentive when the headway starts to shrink, and less aggressive to correct their own vehicle's speed when the headway grows.

Gazis et al. revised the model (2.4) by defining a new sensitivity factor α in (2.3) [24]. They suggest free parameters to be carefully selected based on fitting to measured data. Specifically, driver's sensitivity is formulated as

$$\alpha_k = \frac{\lambda(\dot{x}_k(t))^m}{(x_{k+1}(t-\tau) - x_k(t-\tau))^\ell}, \tag{2.5}$$

where λ, m, and ℓ are constant parameters. Model (2.3) combined with (2.5) by letting $\alpha = \alpha_k$ is called the *non-integer car-following model* [25], which reduces to model (2.4) for $\lambda = \alpha$, $m = 0$, and $\ell = 1$.

Some observations on (2.5) are as follows:

- When $\lambda \neq 0, m = 0$, one should pay attention to the case of when headway approaches zero, which would make α_k arbitrarily large and may not fit well with the physical limitations of a driver's sensitivity to headway variations. Since a smaller headway is analogous to increased traffic density in a macroscopic model, we can state that the model with (2.5) is better suited for low to moderate density levels; see also [26].
- If $\lambda \neq 0$, $m \neq 0$, then the above issue can be resolved, since in this case, α_k can resume a finite limit as vehicles slow down (velocity approaches zero) and traffic density increases (headway approaches zero).

Based on the second item above, it is preferable to select $m \neq 0$ for relatively small velocities and headway. Indeed, traffic flow characteristic will influence the selection of the model parameters. In [27], for example, the authors suggest $m \cong 0.8$ and $\ell \cong 2.8$. See also [28] for further comments on this model.

2.2.3 A Model by Herman and Rothery

One additional feature that can be incorporated into car-following dynamics is the ability of drivers to follow multiple vehicles, including those ahead of or behind them. One such model has been the focus of experimental studies, as reported in [29, 30] and attributed to the work of R. Herman and R.W. Rothery in 1965. This model is a linear delay differential equation where a driver receives stimuli from the two preceding vehicles in the form of velocity errors with a single delay τ, given by

$$\ddot{x}_k(t) = \alpha_1(\dot{x}_{k+1}(t-\tau) - \dot{x}_k(t-\tau)) + \alpha_2(\dot{x}_{k+2}(t-\tau) - \dot{x}_k(t-\tau)), \tag{2.6}$$

with constants α_1 and α_2 representing the sensitivity of the driver against different velocity-error readings. Models similar to this one form the backbone of today's research in the area of connected vehicles that aim to regulate their speeds based on the behavior of multiple vehicles.

2.2.4 A Model by Bando

Since we covered in detail the delay-free version of optimal velocity models in Sect. 1.1.2.1, here we only summarize a model by Bando et al. [2] focused on investigating the effect of time-delays in the nonlinear model

$$\ddot{x}_k(t) = \alpha(V(\Delta x_k(t-\tau)) - \dot{x}_k(t-\tau)), \tag{2.7}$$

with the same parameter definitions as in (1.16). Based on time simulations of (2.7) and a linear stability analysis around the constant-velocity equilibrium point, authors study how different amounts of delay τ could affect the traffic flow characteristics, from homogeneous flow to congestion.

2.2.5 A Model by Konishi

Several other studies consider adding time-delays in the OVM. In 1999, Konishi et al. [31, p. 4001] propose a piecewise linear optimal velocity function with saturation characteristics, instead of a hyperbolic function as in (1.17). This function is then combined with a discrete-time car-following model, whose linear stability around the constant-velocity equilibrium point is studied in the absence of delay effects, and conditions for the stability of traffic flow are obtained.

Moreover, a control scheme is proposed for reducing the traffic jams predicted by the initial model, namely a *decentralized* delayed-feedback controller [31, Eq. (21)]. The proposed controller does not require the velocity information of the other vehicles but only of the vehicle under control. The stability of the controlled vehicle is then studied in order to select the controller gains, followed by simulations with and without the controller, demonstrating that the proposed controller is effective in eliminating traffic jams.

The above-described control scheme is implemented with a sampling period of 0.1 s. This duration, which can be regarded as a delay effect,[4] is about an order of magnitude smaller than the human reaction delay. In [32], a continuous-time model inspired by (2.7) is given, where the car-following dynamics is not influenced by delays, but the feedback controller added to the dynamics is affected by a delay term. The model with this controller reads:

$$\ddot{x}_k(t) = \alpha(V(\Delta_k(t)) - \dot{x}_k(t)) + u_k(t), \tag{2.8}$$

$$u_k(t) = \beta(\Delta_k(t) - \Delta_k(t-\tau)), \tag{2.9}$$

---

[4] Compared to continuous-time problems, the effects of delays can be handled relatively easily in discrete time, as one can define an extended state $\tilde{x}_k$ and a map Φ to describe the dynamics as $\tilde{x}_{k+1} = \Phi\tilde{x}_k$, whose stability is governed by the location of the eigenvalues of Φ with respect to the unit circle.

where $\Delta_k(t) = x_{k+1}(t) - x_k(t)$ is the headway, and the controller $u_k(t)$ needs both the instantaneous headway measurements and the measurements collected τ sec earlier. Further, a linear stability analysis is carried out and the selection of the controller gain β in terms of the delay τ is performed. The stability analysis is performed in the Laplace domain, along with a numerical analysis in the frequency domain, and simulation results are presented for specific choices of (β, τ) pairs.

2.2.6 A Model by Davis

In [33], the author studies collision-free, safe platoon size in the presence of delay. For this, they perform time simulations with 100 vehicles on a car following model that resembles (2.7). Specifically, in their model, the delay effect is included in the optimal velocity function, but not in the term $\dot{x}_k$. Moreover, α in (2.7) is denoted by the time constant $1/\tau$ while delay τ in (2.7) is denoted by t_d in the cited study. The study finds out that even for relatively small amounts of delays, e.g., in the range of 0.3–0.5 s, many vehicles in the platoon collide with each other. Specifically, around delays of 0.2–0.25 s, it is found that the model presents sensitivity in safe platoon size. Considering these discussions, in [16], the author suggests modifications in the optimal velocity function such that the model can handle delays of 0.5–0.1 s. Two modified OVMs are proposed to this end. The first modification suggests adjusting the term $V(\Delta x_k(t))$ in (2.7) as follows:

$$V_{OV} = V(\Delta x_k(t) + \tau \dot{\Delta x_k}(t)), \tag{2.10}$$

and a second modification extends (2.10) to

$$V_{\text{desired}} = \begin{cases} V_{OV} & \text{if } V_{OV} \leq \dot{x}_k(t), \\ \min(V_{OV}, \dot{x}_k(t)) & \text{if } V_{OV} > \dot{x}_k(t). \end{cases} \tag{2.11}$$

In [16], advantages of V_{OV} and V_{desired} are presented based on time simulations of vehicles. It is shown that a platoon of 100 vehicles obeying (2.10) avoids collision for delays up to 1 s; yet this optimal velocity function leads to unrealistically large accelerations/decelerations. This issue is addressed by instead considering (2.11). With (2.11), simulations predict more realistic velocity profiles for the vehicles while not predicting any undesirable levels of oscillations. Moreover, with the proposed modifications, *formation of traffic jams* is observed. However, as stated in [16], under these modifications, the model does not capture *sequence of jam, free-flow phenomena* or *jamming transition*; see [34] for discussions. Further improvements to handle multi-lane traffic and interactions of multi-lane traffic with on-ramps are discussed in [35].

2.2.7 A Model by Treiber

Another key model is the *human driver model* (HDM) [36], which models a human driver in detail considering human reaction times, estimation errors of drivers, as well as temporal and spatial anticipation of drivers when following multiple preceding vehicles. The rationale for implementing driver anticipation is that drivers are aware of their limitations regarding their reaction times; hence, they try to compensate for this weakness by following further ahead of traffic, e.g., by following multiple preceding vehicles. Authors integrate HDM into IDM (see Eq. (1.22)) to study the stability of the arising car-following dynamics, considering a platoon of 100 vehicles. On the parameter plane of driver reaction delay versus the number of vehicles in the platoon, the authors find the parametric regions where traffic flow stability is maintained [36, Fig. 1]. It is shown that the stability regions are considerably enlarged by allowing human drivers to follow the first five vehicles ahead of them rather than only one. These results show that the undesired effects of human reaction times are compensated by an anticipation strategy in the driving model. The authors also identify parameter regions where the traffic exhibits oscillatory behavior or the model predicts a crash.

2.2.8 A Model by Orosz and Stepan

In [37], the authors develop a bifurcation analysis framework in the parameter space of time-delay corresponding to a system with translational symmetry. This mathematical framework is then implemented on a nonlinear microscopic traffic flow model, considering two vehicles traveling around a ring. The vehicles' dynamics are given by

$$\frac{d^2y_1(t)}{dt^2} = \frac{1}{T}\left(\vartheta_0 - \frac{dy_1(t)}{dt}\right) + B(y_2(t-\tau) - y_1(t-\tau)), \tag{2.12}$$

$$\frac{d^2y_2(t)}{dt^2} = \frac{1}{T}\left(\vartheta_0 - \frac{dy_2(t)}{dt}\right) + B(y_1(t-\tau) - y_2(t-\tau) + L), \tag{2.13}$$

where $y_k(t)$ and $\dot{y}_k(t)$ denote, respectively, the positions and velocities of the vehicles, ϑ_0 is the desired velocity, and L is the length of the ring. The nonlinear, continuous function $B(\cdot)$ models braking and is hence negative. Braking is affected by each driver's decision delay τ; in other words, each driver senses the headway between the two vehicles and reacts only after τ sec and as determined by the function B. One can interpret the above dynamics as vehicles aiming to match their velocity $\dot{y}_k$ to the desired velocity ϑ_0 while decelerating as determined by the function B. Via bifurcation analysis, the authors study the characteristics of periodic solutions of the system and the effect of time-delay on such oscillations. One interesting result is that it is possible to mathematically explain *stop-and-go* or *slow-and-go* phenomena—

a widely observed feature of real-world traffic. The analysis also reveals how the particular shape of the braking function B affects these phenomena. Readers are also referred to [38–42] to follow some of the authors' work in the study of traffic flow dynamics and control.

2.3 Notes and Further Reading

The aim of this chapter was to provide a foundation for developing models. The presented models impact today's transportation research, with trends in network interconnections, connected vehicles, and modeling of traffic mixed with human operated and self-driving vehicles. One observation is that most models are of *retarded* type (see, e.g., [43] for a classification of delay differential equations), signifying a common characteristic. Moreover, in most models, the driver follows only one preceding vehicle, although ideas for extending the models for following more than one vehicle are available. Often, human reaction delay is assumed constant to understand the underlying principles. There is evidence that such an assumption can be a good one in relation to experiments [44]. Moreover, behaviors of different drivers may be predicted by different models, and experimental data can be utilized to reveal the best fits [45].

In many studies, a combination of numerical simulations and linear stability around a constant-velocity equilibrium are considered. In the subsequent chapters, we will further investigate this aspect of traffic flow. For studies on nonlinear stability analysis, readers are referred, for example, to [37, 39, 41]. Although delay-free models are easier to handle analytically, large-scale traffic analysis still requires simulation studies, but such simulations may not be scalable.

When studying the effects of human reaction delays, microscopic models are generally more convenient. In such models, it would be of interest to understand the effects of vehicle and delay heterogeneity on traffic flow dynamics, instead of assuming identical vehicle models and delay. Extensions to multiple delays can help capture the influence of possibly different time scales in human decision-making processes, sensing and perception mechanisms, and vehicle-following strategies. For example, it is possible that headway and velocity differences between consecutive vehicles could be sensed with different delays, which may be supported by the remarks in Sect. 2.1.3 that human reaction delays can be different for sensing different acceleration levels of a preceding vehicle.

The presence of delays causes additional challenges in the stability analysis. Multiple, distributed, time-varying delays further exacerbate these challenges. Nevertheless, it remains an inevitable fact that human reaction delays are an integral part of any realistic vehicle driving scenario, which ultimately affect traffic flow dynamics. Hence, it is desirable to understand and utilize appropriate tools for analyzing delayed models. With such tools, an analytical understanding of traffic flow can be established, design strategies can be developed, and automated vehicle-following controllers can be synthesized within a mathematical and computational framework.

Some content in this chapter was reprinted/adapted by permission from Springer Nature: Springer-Verlag, from the chapter *Mathematical models of traffic flow dynamics with time delays* by R. Sipahi and S.-I. Niculescu, published in *Complex Time-Delay Systems: theory and applications* (ed. Fatihcan M. Atay), Copyright 2009 [46]. The presentation here was updated with various references and discussions and completed with insights.

References

1. M. Plöchel, J. Edelmann, Driver models in automobile dynamics application. Veh. Syst. Dyn. **45**, 699–741 (2007)
2. M. Bando, K. Hasebe, K. Nakanishi, A. Nakayama, Analysis of optimal velocity model with explicit delay. Phys. Rev. E **58**, 5429–5435 (1998)
3. M. Bando, K. Hasebe, K. Nakanishi, A. Nakayama, Delay of vehicle motion in traffic dynamics. https://arxiv.org/abs/patt-sol/9608002 (1996)
4. M. Green, "How long does it take to stop?" Methodological analysis of driver perception-brake times. Transp. Hum. Factors **2**(3), 195–216 (2000)
5. M. Treiber, A. Kesting, *Traffic Flow Dynamics: Data, Models and Simulation* (Springer, Berlin, Heidelberg, 2013)
6. K.R. Boff, J.E. Lincoln, Engineering data compendium: human perception and performance, Volume II (Harry G. Armstrong Aerospace Medical Research Laboratory, Wright- Patterson Air Force Base, OH, 1988)
7. W. Schneider, R.M. Schiffrin, Controlled and automatic human information processing: I. Detection, search and attention. Psychol. Rev. **84**(1), 1–66 (1977)
8. W. Schneider, M. Detweiler, The role of practice in dual-task performance: toward workload modeling in a connectionist control architecture. Hum. Factors **30**(5), 539–566 (1988)
9. M. Green, Contrast detection and direction discrimination of drifting gratings. Vision. Res. **23**, 281–289 (1983)
10. M. Green, Visual search, visual streams, and visual architectures. Percept. Psychophys. **50**, 388–403 (1991)
11. M. Green, Visual search: detection, identification, and localization. Perception **21**, 765–777 (1992)
12. M. Green, J. Senders, Human error in road accidents, from the World Wide Web: https://www.visualexpert.com/Resources/roadaccidents.html. Accessed 9 June 2023
13. F.M. Atay, Distributed delays facilitate amplitude death of coupled oscillators. Phys. Rev. Lett.**91**, 094101 (2003)
14. G. Stépán, *Retarded Dynamical Systems: Stability and Characteristic Function* (Longman Scientific & Technical, Wiley, US, New York, 1989)
15. R. Sipahi, F.M. Atay, S.-I. Niculescu, Stability of traffic flow with distributed delays modeling the memory effects of the drivers. SIAM Appl. Math. **68**(3), 738–759 (2008)
16. L.C. Davis, Modifications of the optimal velocity traffic model to include delay due to driver reaction time. Phys. A **319**, 557–567 (2003)
17. R.W. Allen, J.D. Marcotte, T.J. Rosenthal, B.L. Aponso, Driver assessment with measures of continuous control behavior, in *Driving Assessment Conference*, vol. 3, no. 2005. University of Iowa (2005)
18. R.E. Chandler, R. Herman, E.W. Montroll, Traffic dynamics: studies in car following. Oper. Res. Int. J. **6**(2), 165–184 (1958)
19. A. Reuschel, Fahrzeugbewegungen in der Kolonne. Oesterreichisches Ingenieur-Archiv **4**, 193215 (1950)
20. A. Reuschel, Z. Oster, Ingenieur - Arch. Vereines **95**, 59 (1950)

21. J. Zhang, P. Ioannou, Control of heavy-duty trucks: environmental and fuel economy considerations (University of Southern California, Los Angeles TO4203, 2004)
22. L.A. Pipes, An operational analysis of traffic dynamics. J. Appl. Phys. **24**(3), 274–281 (1953)
23. D.C. Gazis, R. Herman, R.B. Potts, Car-following theory of steady-state traffic flow. Oper. Res. **7**(4), 499–505 (1959)
24. D.C. Gazis, R. Herman, R.W. Rothery, Nonlinear follow-the-leader models of traffic flow. Oper. Res. **9**(4), 545–567 (1961)
25. D. Helbing, Traffic and related self-driven many-particle systems. Rev. Mod. Phys. **73**, 1067–1141 (2001)
26. W.F. Phillips, A new continuum traffic model obtained from kinetic theory, in *1978 IEEE Conference on Decision and Control, San Diego, CA* (1979), pp. 1032–1036
27. Jr. A.D. May, H.E.M. Keller, Non-integer car-following models, Highway Research Record, No. 199, (1967), pp. 19–32
28. D. Helbing, B. Tilch, Generalized force model of traffic dynamics. Phys. Rev. E **58**, 133–138 (1998)
29. Traffic flow theory—a state of the art report, ed. by N.H. Gartner, C.J. Messer, A.K. Rathi (2001)
30. D.L. Gerlough, M.J. Huber, *Traffic Flow Theory*, Transportation Research Board (TRB) Special Report 165 (Washington D.C., 1975)
31. K. Konishi, H. Kokame, K. Hirata, Coupled map car-following model and its delayed-feedback control. Phys. Rev. E **60**, 4000–4007 (1999)
32. K. Konishi, H. Kokame, K. Hirata, Decentralized delayed-feedback control of an optimal velocity traffic model. Eur. Phys. J. B **15**, 715–722 (2000)
33. L.C. Davis, Comment on "analysis of optimal velocity model with explicit delay". Phys. Rev. E **66**, 038101 (2002)
34. N. Mitarai, H. Nakanishi, Spatiotemporal structure of traffic flow in a system with an open boundary. Phys. Rev. Lett. **85**(8), 1766–1769 (2000)
35. L.C. Davis, Multilane simulations of traffic phases. Phys. Rev. E **69**, 016108 (2004)
36. M. Treiber, A. Kesting, D. Helbing, Delays, inaccuracies and anticipation in microscopic traffic models. Phys. A **360**(1), 71–88 (2006)
37. G. Orosz, G. Stépán, Hopf bifurcation calculations in delayed systems with translational symmetry. J. Nonlinear Sci. **14**(6), 505–528 (2004)
38. G. Orosz, R.E. Wilson, R. Szalai, G. Stépán, Exciting traffic jams: nonlinear phenomena behind traffic jam formation on highways. Phys. Rev. E **80**, 046205 (2009)
39. G. Orosz, R.E. Wilson, B. Krauskopf, Global bifurcation investigation of an optimal velocity traffic model with driver reaction time. Phys. Rev. E **70**, 026207 (2004)
40. G. Orosz, B. Krauskopf, R.E. Wilson, Bifurcations and multiple traffic jams in a car-following model with reaction time delay. Phys. D **211**, 277–293 (2005)
41. G. Orosz, R.E. Wilson, B. Krauskopf, Bifurcations in a car-following model with delay, in *IFAC Workshop on Time Delay Systems 2004*. Leuven, Belgium
42. G. Orosz, E. Wilson, G. Stépán (eds.) Traffic jams—dynamics and control. Philos. Trans. R. Soc. A **368**(1928) (2010)
43. J.K. Hale, S.M. Verduyn Lunel, *Introduction to Functional Differential Equations* (Springer, New York, 1993)
44. A. Bose, P.A. Ioannou, Analysis of traffic flow with mixed manual and semi automated vehicles. IEEE Trans. Intell. Transp. Syst. **4**(4), 173–188 (2003)
45. C.P.I.J. Van Hinsbergen, W.J. Schakel, V.L. Knoop, J.W.C. Van Lint, S.P. Hoogendoorn, A general framework for calibrating and comparing car-following models. Transp. A: Transp. Sci. **11**(5), 420–440 (2015)
46. R. Sipahi, S.-I. Niculescu, Mathematical models of traffic flow dynamics with time delays, in *Complex Time-Delay Systems: theory and applications*. ed. by Fatihcan M. Atay (Springer, 2010)

Chapter 3
Stability Analysis Framework of Delay Systems

In this chapter, we first provide some motivations on the importance of time-delays in dynamical systems. More precisely, we show how time-delays can have intriguing and at times counter-intuitive effects on the behavior of dynamical systems. This discussion is followed by the definitions of linear stability and the spectrum of time-delay systems, together with two key notions, namely D- and tau-partitioning methods, with which stability is studied with respect to partitions in the space of time-delays and system parameters. The stability analysis and the relevant tools are summarized with illustrative examples to demonstrate the main ideas. The focus of the presentation of this chapter is on stability, while the definition of slinky effects is left to Chap. 6.

3.1 Dual Effects of Delays, Definitions, and Main Concepts

3.1.1 Motivation

Time-delays exist in many real-world dynamical systems, including both human-built engineering systems and natural systems [1–3]. Indeed, one may even say that delays are everywhere. For instance, in tele-operation, a communication medium is used to transmit signals from one site to a remote site. Such a medium inevitably causes delays in signal transmission, which can be large enough to affect the performance of the tele-operation process. In supply chains, product transport delays are analogous to transmission delays, which retard the arrival of new product shipments to inventories [4]. In biological systems and population dynamics, delays may be in the form of maturation times and are related to hereditary after-effects [5, 6]. Other examples are found in human motor control problems. For instance, modeling human posture control requires the consideration of about 80 ms neurological delays between the central nervous system and muscle activation at human ankle [7].

R. Sipahi et al., *Deterministic Car-Following Traffic Models*,
SpringerBriefs in Control, Automation and Robotics,
https://doi.org/10.1007/978-3-031-58164-9_3

As delays appear all around us, it is natural to ask *how delays affect dynamical behavior*. Intuitively, considering a tele-operation system with delays, the remote robot would produce actions based on delayed information received from the other site. However, if the controller of this robot is designed only to handle instantaneous information, then the actions calculated by the controller would not be appropriate when faced with delayed information; that is, those actions would not have the capability to compensate against current loads and disturbances. Consequently, the control actions produced would not be the most appropriate, and hence the entire tele-operation process may perform poorly, or become "unstable" as characterized by oscillatory behavior or growing deviations from an intended behavior.

Delays do not always lead to instability—a somewhat counter-intuitive outcome [1]. For instance, in many studies, it was shown that larger delays, or adding intentional delays in decision-making, do not always indicate poorer results on dynamic behavior [8, 9]. A physical example of this is the famous water-temperature control in the shower, which is inherently time-delayed since a change in the knob can be felt only after a finite time. Turning the knob rapidly without waiting to feel the changes in temperature is not desired due to this delay as such actions can easily lead to undesirable temperature fluctuations. However, if the user adds more delays in the decision-making, e.g., by waiting for some time to sense the temperature variation before re-adjusting the knob, the water temperature can be effectively regulated.

- It is commonly recognized that time-delays can be a source of instability and poor performance. However, in some cases delays may actually induce stability and may represent a simple way to improve the dynamical behavior. For example, although an oscillator cannot be stabilized with a proportional control alone, it can be stabilized by adding a small delay in the closed loop [10]. Hence, it is important to have a finer characterization of the effects induced by the delays on system's stability.
- In some cases, delays can be treated as "design" (or control) parameters. If such a design is possible, this would offer more flexibility for selecting these delays for the system in order to guarantee stability [9, 11] and good performance [12, 13]. For instance, a chain of n integrators cannot be stabilized using one delay block, but the closed-loop stability can be achieved by using some appropriate n distinct delay blocks [14].

Without an appropriate analytical framework, it is not straightforward to estimate how the dynamics may be affected in the presence of delays. A rigorous stability analysis is essential to answer this question; however, such an analysis in the presence of time-delays poses challenges beyond those found in problems without delays and becomes even more complicated when multiple delays are present [15]. Recent advances in this direction from different perspectives [16, 17] can be useful for studying traffic flow models in which multiple delays may arise from multiple human drivers or sensing systems. Moreover, results from control systems with delays could be revisited to study the existing models; see, for instance, [18–21]. These efforts can be extended toward understanding automated headway controllers, and collaborative vehicles where communication delays arise naturally as part of the network architecture [22–25].

As motivated above, the dual nature of delays, e.g., their stability- or instability-inducing characteristics, calls for the development of a mathematical foundation by which such characteristics can be identified and various scenarios of dynamical behavior can be studied, understood, and possibly reshaped according to the control objective at hand. At this point, many opportunities are apparent in the analysis and design of traffic flow dynamics. To present these systematically and to point out future directions, we start by giving a formal and intuitive mathematical description of the concepts, discussing the key points in the stability analysis, and presenting numerical toolboxes with which reliable results can be obtained. Several examples round up the presentation and are useful to better understand the underlying ideas, concepts, and methods.

3.1.2 *Definitions and Prerequisites*

We mostly follow the mathematical formalism from [26, 27]. Let τ denote the maximum positive delay in the system, and let $X = \mathscr{C}([-\tau, 0], \mathbb{R}^n)$ denote the Banach space of continuous functions mapping the interval $[-\tau, 0]$ to $\mathbb{R}^n$, with norm defined as $\|\phi\| = \sup_{-\tau\leq\theta\leq 0} |\phi(\theta)|_\infty$ where $|\cdot|_\infty$ is the infinity norm on $\mathbb{R}^n$. We will mainly focus on autonomous delay differential equations, that is, equations of the form

$$x'(t) = F(x_t), \quad t \geq 0, \tag{3.1}$$

where $()'$ denotes the right-hand derivative, the delayed state $x_t \in X$ is defined as $x_t(\theta) := x(t+\theta)$ for $\theta \in [-\tau, 0]$, and an initial condition x_0 is specified as $x_0 = \phi$ for some $\phi \in X$. For a deeper discussion on the initial-value problem for the differential equation above, the reader is referred to [28]. Since the initial condition ϕ as well as the state x_t are functions in X, such differential equations, generically called *functional differential equations*, are not finite-dimensional. Moreover, the particular Eq. (3.1) is said to be of *retarded* type since the rate of change of the current state does not depend on past values. For further classifications and properties, see, for instance, [28, 29] and the references therein.

When the autonomous system (3.1) is linear, it can be written as

$$x'(t) = Lx_t, \quad t \in \mathbb{R}, \tag{3.2}$$

where $L : X \to \mathbb{R}^n$ is a linear operator. This class of systems can be expressed more explicitly in the form

$$x'(t) = \sum_{i=0}^{N_p} A_i x(t-\tau_i) + \sum_{\mu=1}^{N_d} \int_{-\tau_\mu}^{-\tau_{\mu-1}} \Gamma_\mu(\theta) x(t+\theta)\, d\theta, \tag{3.3}$$

where $x(t) \in \mathbb{R}^n$ is a vector, $N_p, N_d \in \mathbb{N}$, the $A_i \in \mathbb{R}^{n\times n}$ are matrices with constant entries, and the kernel $\Gamma_\mu(\theta)$ has entries continuous with respect to θ, and for any compact interval in $\mathbb{R}$, there exists a $\gamma_\mu \in \mathscr{C}([-\tau_\mu, -\tau_{\mu-1}], \mathbb{R})$ such that the kernel is norm bounded, $\|\Gamma_\mu(\theta)\| \leq \gamma_\mu(\theta)$. Moreover, the distinct delays τ_ℓ satisfy, without loss of generality, $0 = \tau_0 < \tau_1 < \cdots < \tau_{N_p} := \tau$.

Although systems of the form (3.3) are infinite-dimensional, they exhibit certain continuity properties regarding their spectrum, resembling those of ordinary differential equations. The delays under the first summation in (3.3) are called *discrete* (point-wise) delays since the effects arrive from a point in time in the past, while the delays arising in the integral terms are called *distributed* delays, as their effects arrive from a continuum of points from the past weighted by $\Gamma_\mu(\theta)$. If $N_d = 0$, that is, if (3.3) includes only discrete delays, it is sometimes called a *differential-difference equation* [30] or *retarded differential-difference equation* [28].

3.2 Asymptotic Stability

In linear time-invariant systems, asymptotic stability implies exponential stability. In this book, we will call asymptotically stable systems simply "stable" for brevity. There exists an abundant literature on the stability analysis of delay systems; see, e.g., [1, 29, 31, 32] for existing approaches and methodologies. In this chapter, we mainly focus on frequency-domain techniques and present guidelines for the stability analysis of linear time-invariant systems with constant delays.

3.2.1 *Characteristic Roots, Root Continuity, and D-Partitioning*

We present several key concepts regarding the spectrum of (3.3) that will be essential for the stability analysis. Without losing the essence of the discussion, we will consider only discrete delays, referring the readers to [20] on the spectrum of delay differential equations with distributed delays.

In the case of discrete delays, the characteristic function can be interpreted as a function of the delay parameters $f : \mathbb{C} \times \mathbb{R}^{N_p} \mapsto \mathbb{C}$. By taking the Laplace transform of (3.3) or by considering exponential solutions of the form $x(t) = e^{st}\mathbf{b}$ for some constant vector $\mathbf{b}$, one obtains the characteristic function:

$$f(s; \tau_1, \ldots, \tau_N) := \det\left[sI - A_0 - \sum_{i=1}^{N} A_i e^{-s\tau_i}\right], \tag{3.4}$$

where I is the $n \times n$ identity matrix. The function f is an entire function of special type, called a *quasi-polynomial* (see, e.g., [30, p. 393]). It possesses an infinite number of zeros $s \in \mathbb{C}$, called the *characteristic roots*. As in the finite-dimensional case, the location of these roots in the complex plane determines the stability of the system.

More precisely, for a given set of delays, the zero solution is asymptotically stable if and only if all of the roots of (3.4) lie in the open left-half complex plane $\mathbb{C}_-$. Verifying asymptotic stability is not a straightforward task since (3.4) has infinitely many characteristic roots. To investigate asymptotic stability, one should first leverage the continuity property of the spectrum [20, 33].

We have that the roots of a polynomial are continuous functions of coefficients, with a non-vanishing leading coefficient [34]. Furthermore, polynomials are differentiable in the case of simple roots. For multiple roots, differentiability is lost, and the Puiseux series can be used to analyze such cases [34, 35]. These ideas are also valid for linear delay systems of retarded type. Since our focus is on the stability effects induced by the delays, seen as parameters, it is natural to define a function $g : \mathbb{R}_+^{N_p} \mapsto \mathbb{R}$ by

$$g(\tau_1, \ldots, \tau_{N_p}) := \sup \left\{ \Re(s) : \ f(s; \tau_1, \ldots, \tau_{N_p}) = 0, \ s \in \mathbb{C} \right\}, \tag{3.5}$$

called the *spectral abscissa* function, which defines the real part of the rightmost characteristic root(s) as a function of the delays. In this case, stability analysis reduces to checking the sign of $g(\cdot)$. We note that g *always exists* and is *finite* for differential-difference equations. Furthermore, $g(\cdot)$ is a *continuous function* of both τ_k and system's parameters [29, 33, 36]. The continuity property suggests having a deeper look at three important problems: (i) analyzing the stability of the system without delays, (ii) studying the existence of critical characteristic roots located on the imaginary axis, and (iii) analyzing the asymptotic behavior of such critical roots with respect to the considered (delay) parameters.

A particular case study is represented by systems that do not have critical characteristic roots on the imaginary axis for any value of the delays. A linear delay system of retarded type is called *hyperbolic* at some $\tau = (\tau_1, \ldots, \tau_{N_p}) \in \mathbb{R}_+^{N_p}$ if $\Re(\lambda) \neq 0$ for any zero λ of f in (3.4). If the system is hyperbolic for all delay values, then the system is *delay-independent hyperbolic* (see, e.g., [37, 38] for further details and characterizations). Although quite restrictive, such systems have an interesting property: in certain problems, stability of a delay-independent hyperbolic system can be directly deduced from the stability of the delay-free system.

In order to fix better the underlying ideas, consider a simpler case with the characteristic function (3.4) of the form:

$$f(s; \tau) := P_0(s) + P_1(s)e^{-s\tau}, \tag{3.6}$$

where P_0 and P_1 are polynomials in s. Then, the corresponding system is *delay-independent hyperbolic* if

(i) the function f does not possess a root at the origin, i.e., $P_0(0) + P_1(0) \neq 0$, and
(ii) the sweeping condition $|P_1(i\omega)| < |P_0(i\omega)|$ holds for all $\omega > 0$.

Indeed, the property follows in a straightforward fashion from the fact that $|e^{-i\omega\tau}| = 1$, for all $\omega \in \mathbb{R}$ and $|f(0; \tau)| > 0$ for all $\tau \in \mathbb{R}_+$. Finally, it is worth mentioning that the conditions (i)–(ii) above are *necessary and sufficient* for f defined in (3.6) to be delay-independent hyperbolic.

Remark 3.1 (*Scalar case*) In order to better illustrate the property above, consider now the scalar case in (3.6): $P_0(s) = s + a$, $P_1(s) = b$, with $a, b \in \mathbb{R}$ and $b \neq 0$. Some simple computations show that the *delay-independent hyperbolicity* property is equivalent to $a + b \neq 0$, and $|a| \geq |b|$. Furthermore, if $a + b > 0$ (or, $a + b < 0$), then the zero solution of the corresponding scalar equation is *delay-independent* stable (or, unstable). Finally, it is worth mentioning that if $a + b < 0$, $|a| < |b|$, the scalar system is *delay-independent* unstable without being delay-independent hyperbolic.

Suppose now that the delay-free system is hyperbolic, but the delay system is not delay-independent hyperbolic. In this case, the continuity of the spectral abscissa function guarantees that the stability/instability of the delay-free system is conserved for sufficiently small delay parameters. Furthermore, there are no roots coming from $+\infty$ in the right-half plane $\mathbb{C}_+$, when the delay parameters increase from 0 to 0^+. Next, under the assumption that for some $k \in \{1, \ldots, N\}$, the delay τ_k is a free parameter and all the other delays τ_i $(i \neq k)$ are fixed, the system can "switch" from stability to instability, or vice versa, only when characteristic roots "touch or cross" the imaginary axis as τ_k changes. In other words, the stability analysis of the system requires calculating (if any) the characteristic roots located on the imaginary axis $s = i\omega$, to explicitly analyze their asymptotic behavior with respect to the corresponding delay parameter. This approach is the basis of most of the frequency-domain-based stability analysis techniques developed in the literature; see [19, 20, 39] as well as some applications in [3, 40].

Assume now that the delay system (3.3) has only one constant delay τ and that the characteristic function f has a pair of imaginary roots at $s = \pm i\,\omega^*$ for some delay value $\tau = \tau^*$ and all the remaining roots have negative real parts. In other words, the rightmost root is located on the imaginary axis, i.e., $g(\tau^*) = 0$. Let us assume that as we increase τ past the value τ^*, the roots at $\pm i\,\omega^*$ migrate into the right-half complex plane. As we keep increasing τ, one possibility is that these roots may turn around and, for some $\tau = \tilde{\tau} > \tau^*$, they cross over the imaginary axis at $s = \pm i\,\tilde{\omega}$. While this is happening, assume that all the remaining roots still remain in the left-half complex plane. In this setting, the delays τ^* and $\tilde{\tau}$ define the critical values of a delay interval in which the system is always unstable.

Expanding on this example, we have that the system's stability/instability will remain the same within each such delay interval whose boundaries are determined by critical delays for which the system has roots on the imaginary axis. Inspired by the D-partitioning method[1] [41], this partition of the (positive) delay axis (known as *tau-partitioning* [42]) allows to divide the delay axis into several intervals such that the number of unstable roots is the same for all the delay values within an interval.

[1] In the particular case of only one "free" parameter.

Intuitively, such an idea can be extended to handle *multiple delays* [17, 43]. In [17], for example, the authors demonstrate how to obtain the stability crossing curves for a two-delay problem and reveal the stability/instability decomposition of the system on the delay-parameter plane, while in [16, 44], the authors provide a classification of the stability crossing curves. It is worth mentioning that the stability crossing curves consist of all the delay values such that the characteristic function has at least one root on the imaginary axis. In the case of more than two delays, the parameter regions where this number is fixed are found in high-dimensional volumes that are separated from each other by hypersurfaces.

In the discussion above, we have assumed that the delay-free system is *hyperbolic*. If such a property does not hold, then two situations may appear: (i) $s = 0$ is a root of the corresponding characteristic function $f(s; 0, \ldots 0) = 0$, and (ii) all the imaginary roots of f are not located at the origin.

In order to have a better understanding, recall $f(s; \tau)$ of the simpler case (3.6), which includes only one discrete delay:

In (i), if 0 is a characteristic root of the delay-free system, that is, $f(0; 0) = 0$, then it is easy to see that $f(0; \tau) = 0$. In other words, such a root is *invariant* for all delay values, and varying the delay will not "move" it from the origin. This scenario will be revisited in Chap. 5. In the scalar case (see, for instance, Remark 3.1), the root at the origin corresponds to the case when $a + b = 0$. Furthermore, if $a\tau = 1$, it is easy to see that this root is double, which corresponds to the maximum multiplicity of an imaginary root for this system.

In the second case (ii), as the delay parameter changes, we intuitively expect that the imaginary roots will "move" away from the imaginary axis either to the left or right-half complex plane. It is worth mentioning that, in the scalar case, the movement is only to the right, but for higher-dimensional systems, such movements can also be to the left. Although such a discussion is out of the scope of this brief, the case of an oscillatory second-order system subject to a single discrete delay in the input/output channel is briefly discussed in the sequel; see, for instance, Sect. 3.2.4 below.

3.2.2 *Essentials of Linear Stability Analysis*

For the simplicity of argument, we will explain the concepts of stability analysis in a one-delay-parameter space assuming that all other (system) parameters are given and fixed.

As discussed above, when the delay is zero $\tau = 0$, we have a delay-free system, i.e., the standard linear stability problem for finite-dimensional systems, which is trivial to address. For the sake of simplicity, assume now that the delay-free system is hyperbolic and that we know the number of unstable roots of the corresponding characteristic function. Assume further that the delay system is not delay-independent hyperbolic.

In other words, in this configuration, there exists some delay value τ^* for which the characteristic function has at least one root on the imaginary axis. Based on the continuity properties of the spectral abscissa, it follows that a change in the system's stability is related to the existence of characteristic roots on the imaginary axis, the so-called critical roots or eigenvalues of the system, and on the asymptotic behavior of these critical roots with respect to the corresponding delay parameter:

i. **Imaginary Axis Crossings of Critical Eigenvalues**: Various techniques can be utilized to compute the critical eigenvalues $s^* = \pm i\omega^*$ of a LTI (linear time-invariant) time-delay system and its corresponding delay values τ^*; see, e.g., [5, 17, 31, 38, 45–47]. By the periodicity of trigonometric functions, $\tau_k = \tau^* \pm 2\pi/\omega^*$ also yields the same critical eigenvalue s^*. In general, all the techniques reduce the computation to an equivalent eigenvalue problem, although in many cases this connection may not be explicit. Some are based on a parametric study, e.g., the *bilinear transformation* founded in Rekasius's work [46], while some require *parameter sweeping*, e.g., *frequency sweeping* [48]. Moreover, elegant *elimination techniques* were utilized with various ideas; see [38, 49, 50]. Other studies can be found in [18, 20, 29, 31].

ii. **Sensitivity Analysis of Critical Eigenvalues**: Once the critical eigenvalues $s^* = \pm i\omega^*$ and τ^* are computed, it is important to study in what direction s^* tends to move on the complex plane as the delay τ is slightly increased from τ^*. Since the direction of the real part of s^* matters from the stability point of view, one can use the characteristic function (3.4) to compute this direction. Assuming that this is a simple crossing across the imaginary axis one can use the implicit function theorem to write

$$df = 0 \quad \Longrightarrow \quad \frac{\partial f}{\partial s} ds + \frac{\partial f}{\partial \tau} d\tau = 0, \tag{3.7}$$

which can be re-arranged as

$$\mathscr{S}(s, \tau) := \frac{ds}{d\tau} = -\left(\frac{\partial f}{\partial s}\right)^{-1} \frac{\partial f}{\partial \tau}, \tag{3.8}$$

and the horizontal movement of s^* at τ_k^* is calculated from $\Re(\mathscr{S}(s^*, \tau_k^*))$. If $\Re(\mathscr{S}) > 0$ (resp., < 0) then s^* moves to the right-half (resp., left-half) complex plane as the delay parameter is increased past τ_k^*. This information enables one to keep an account of the number of unstable roots of the system as the parameter τ_k crosses the boundary τ_k^* and moves from one parameter region to the neighboring one [36]. Relevant studies along these lines can also be found in the work cited above in (i).

Remark 3.2 The framework outlined above presents the key points and the general approach to stability analysis. This information is distilled from numerous publications cited and is presented here concisely to explain the main spirit of the approach. Further discussions on individual techniques, their strengths and weaknesses, their capabilities, and computational complexities are left outside the discussion here, as they do not serve the scope of this book. The case studies presented below will provide further insight into only some of the approaches, due to limited space.

3.2.3 *Stability Maps*

When studying stability on the system characteristic equation (3.4), one of the main objectives is to visualize the parametric region(s) where system stability holds. This visualization, as supported by root continuity and D- or tau-partitioning principles, provides useful information about how system parameters could affect stability, and how far the chosen parameters are from the boundary of the region (non-fragility concept). Since in the case of delay-independent stability, visualization of stability does not offer additional information, published work is focused on delay-dependent stability and how system parameter space is decomposed into stable and unstable regions.

When all the delays τ in the dynamics are identical to each other, we are then dealing with a system with a *single delay*, that is, $N_p = 1$ in (3.4). The stability of such systems can be analyzed and portrayed along this *delay parameter*. For example, the system may exhibit stable behavior in a single-delay interval, or it may happen that stability exists in multiple disjoint delay intervals [19, 51]. In the presence of two different delays, τ_1 and τ_2, one can extend the idea of intervals for single delays to instead displaying stability on the 2D plane, known as *stability charts* [20]. Such charts present for which combinations of delays the system is stable or not, and they can also be created on the plane of a delay versus a controller gain to allow interpreting how the gain must be selected for a given delay such that system stability is maintained.

A stability chart will also facilitate the judgment as to whether increasing a delay value favors stability or instability. Moreover, it may turn out that the dynamics may exhibit stability regardless of how large the delays are (delay-independent stability) so long as one remains on some special fixed ratios τ_2/τ_1 between the two delays. Such characteristics may however be lost under small perturbations of this ratio, yielding multiple switches between stability and instability. This phenomenon is especially critical in understanding the robustness of controllers designed for time-delay systems. Some ideas on stable delay rays as well as related delay-interference phenomena are presented in Sect. 3.3.

Depicting stability charts in dimensions higher than three is challenging, not only due to visualization constraints but also due to increased computation times needed to calculate the critical eigenvalues of (3.3). This is, in general, a nondeterministic polynomial NP-hard problem for $N_p > 1$ [15]. In this case, NP-hard complexity reflects itself as intensive computational times needed to solve an example of a decision question as a function of the size of its input. NP-hard problems are computationally costly in this setting; however, some remedies to reduce computation times have recently been proposed for multiple-delay problems [1].

3.2.4 *Illustrative Numerical Examples*

To illustrate the stability analysis on a concrete example, we consider the delay differential equation (DDE)

$$x'(t) = -kx(t-\tau),$$

where k is a nonzero real number. By taking the Laplace transform of both sides, the corresponding characteristic equation is found to be

$$f(s;\tau) := s + ke^{-s\tau} = 0.$$

In control engineering language, the above equation represents a closed-loop system with a plant whose transfer function is given by $H(s) = 1/s$ and a controller in the form of $C(s) = -ke^{-s\tau}$. Here, τ is the delay in the feedback, and k is the controller gain.

When $\tau = 0$, the system is hyperbolic since $k \neq 0$. More precisely, the characteristic function f has a single root at $s = -k$, located on the real axis. As τ is increased continuously from zero to 0^+, the root that was at $s = -k$ moves infinitesimally in $\mathbb{C}$, while at the same time an infinite number of roots $s = \tilde{s}_q, q = 1, 2, \ldots$, appear in $\mathbb{C}$. These infinitely many roots satisfy two conditions, namely $\Re(\tilde{s}_q) < 0$, and $|\tilde{s}_q| \to \infty$, as $\tau \to 0^+$. That is, for a sufficiently small delay, the roots $\tilde{s}_q$ do not affect system stability. On the other hand, when the delay parameter is further increased, the real parts of these roots may increase, and consequently some of these roots can cross over the imaginary axis toward the right-half plane, thus *destabilizing* the system.

3.2.4.1 Delay as a Destabilizing Parameter

Consider the above example with $k > 0$ [1]. To study the closed-loop system stability, one starts with checking if the characteristic roots can ever touch the imaginary axis. For this, setting $s = i\omega$ in the closed-loop characteristic equation, we have

$$s + ke^{-s\tau} = 0, \tag{3.9}$$

for all τ, that is,

$$\cos(\omega\tau) = 0, \quad k\sin(\omega\tau) = \omega. \tag{3.10}$$

Since $k \neq 0$, it is easy to see that 0 is not a characteristic root of f, and thus $\omega \neq 0$. Furthermore, due to the periodicity property of trigonometric functions in (3.10), there exist infinitely many delays $\tau_{c,\ell} = \pi/(2k) + (2\pi\ell)/\omega_c$, $\ell = 0, 1, 2, \ldots$ (the subscript "c" denoting "crossing"), all of which yield the crossing frequency[2] $\omega_c = k$. In other words, Eq. (3.9) has roots on the imaginary axis at $s = \pm ik$. By continuity argument, it then follows that closed-loop stability is guaranteed for all delays satisfying $\tau \in [0, \tau_c)$, where $\tau_c = \pi/(2k)$ is the smallest positive delay for which the crossing occurs. In this example, the system is unstable for $\tau \geq \tau_c$, and thus τ_c is the *delay margin* of the system.

The number of unstable roots can also be determined with respect to the delay parameter τ. For this, the crossing direction of an imaginary root as a function of τ should be evaluated at the corresponding crossing frequency ω_c. Since the quantity $\Re\{ds/d\tau\}|_{s=\omega_c i} = \frac{\omega_c^2}{1+\omega_c^2\tau^2}$ is positive in this example, it indicates that an increase of the delay past each critical delay value $\tau = \tau_{c,\ell}$ corresponds to the crossing of the imaginary axis by a pair of characteristic roots toward $\mathbb{C}_+$. The number of unstable roots NU can then be tracked as a function of the delay parameter. The behavior of the characteristic roots is investigated by using perturbation-based analysis in [29, 52] and following root clustering techniques in [17, 43].

Let us now utilize the bilinear transformation proposed by Rekasius [46]:

$$e^{-\tau\omega i} := \frac{1 - T\omega i}{1 + T\omega i}, \tag{3.11}$$

where $T \in \mathbb{R}$ is a parameter introduced to facilitate the calculations. With the above transformation, exponential functions are replaced by the right-hand bilinear transformation converting the original characteristic equation to a new polynomial form in terms of T. Note that this transformation is *not* a Padé approximation since T is a free parameter.[3] Moreover, since $T \in \mathbb{R}$, we have that $1 + T\omega i \neq 0$; hence, this term can be used to remove the fractions by expanding the arising polynomial, without introducing any artificial imaginary roots.

Under the Rekasius transformation, the characteristic equation (3.9), with $s = i\omega$, $\omega \neq 0$, and with an expansion by $(1 + T\omega i)$, leads to $i\omega(1 + T\omega i) + k(1 - T\omega i) = 0$, which can be re-arranged as

$$-T\omega^2 + k + (\omega - kT\omega)i = 0. \tag{3.12}$$

Given k, if there exists $T \in \mathbb{R}$ such that the above equation is satisfied for some ω, then ω is a critical crossing of the system. If so, one can plug in T and ω solutions into (3.11) to back-calculate the critical delay τ that gives rise to the critical crossing. Calculation of delay can be performed based on the phase conditions of (3.11),

[2] Since the characteristic roots are symmetric with respect to the real axis of the complex plane and that $f(0, \tau) \neq 0$, it suffices to consider only $\omega > 0$.

[3] A first-order Padé approximation would fix $T = \tau/2$.

$$\tau_\ell = \frac{2}{\omega}\left(\tan^{-1}(\omega T) \mp \ell\pi\right), \tag{3.13}$$

$\ell = 0, 1, 2, \ldots$, and the pair $(\omega, T) \in \mathbb{R}_+ \times \mathbb{R}$ satisfies (3.12).

Let us now solve for the root pairs $(\omega, T) \in \mathbb{R}_+ \times \mathbb{R}$ satisfying (3.12). This can be done simply by simultaneously solving the common roots of the real and imaginary parts of (3.12). For $k \neq 0$, we find $T = 1/k$ and $\omega = k$; hence, from (3.13),

$$\tau_\ell = \frac{2}{k}\left(\frac{\pi}{4} \mp \ell\pi\right) = \frac{\pi}{2k} \mp \frac{2\pi\ell}{\omega}, \tag{3.14}$$

which is consistent with the derivation just below (3.10).

We close this subsection by referring interested readers to the discussions in Sect. 3.2.2 regarding the computation of critical characteristic roots of LTI time-delay systems, as well as theoretical developments on root clustering techniques in [17, 43, 45].

3.2.4.2 Delay as a Stabilizing Parameter

As a second example, consider the second-order delay differential equation

$$x''(t) + \omega_0^2 x(t) = kx(t-\tau),$$

whose characteristic equation is

$$f(s;\tau) := s^2 + \omega_0^2 - ke^{-s\tau} = 0. \tag{3.15}$$

This can be viewed as a system with open-loop transfer function $H(s) = 1/(s^2 + \omega_0^2)$ in feedback with the delayed controller $C(s) = ke^{-\tau s}$ [10].

Some simple computations show that the delay-free system is hyperbolic for $|k| > \omega_0^2$, whereas it is not hyperbolic for $|k| < \omega_0^2$ with imaginary roots $\pm i\sqrt{\omega_0^2 - k}$. Finally, for $k = \omega_0^2$, there is a double root at the origin.

Consider now the case of a strictly positive k, such that $k \in (0, \omega_0^2)$.

In this case, following an analysis similar to that for (3.9), we can show that two distinct crossing frequencies exist, $\omega_{c,1} = \sqrt{\omega_0^2 - k}$ and $\omega_{c,2} = \sqrt{\omega_0^2 + k}$. Each crossing is created by infinitely many critical delay values, respectively, $\tau_{c,1,\ell} = (2\ell\pi)/\sqrt{\omega_0^2 - k}$ and $\tau_{c,2,\ell} = (2\ell + 1)\pi/\sqrt{\omega_0^2 + k}$, $\ell = 0, 1, 2, \ldots$. Moreover, in view of (3.15), the sensitivity expression

$$\Re\left\{\frac{ds}{d\tau}\right\}\bigg|_{s=i\omega_c} = -\frac{(\omega_0^2 - \omega_c^2)\omega_c^2}{\omega_c^2 + \tau^2(\omega_0^2 - \omega_c^2)^2}$$

can be simplified to sgn $\Re\{ds/d\tau\}|_{s=\omega_c i} = \text{sgn}(\omega_c^2 - \omega_0^2)$, which indicates that the characteristic roots crossing at $\omega_c = \omega_{c,1}$ favor stability as per negative sensitivity, that is, the roots move toward $\mathbb{C}_-$ across $s = \pm i\omega_{c,1}$, whereas the roots crossing at $\omega_{c,2}$ favor instability as per positive sensitivity.

When $\tau = 0$, the closed-loop system has only a pair of poles at $s = \pm i\,\omega_{c,1}$. These poles are created by the delay values $\tau_{c,1,\ell}$, and at these delay values, pole crossings favor stability. For this reason, when the delay is slightly increased from zero, $\tau = \varepsilon > 0$, the closed-loop system will stabilize as the poles $s = \pm i\,\omega_{c,1}$ migrate to $\mathbb{C}_-$ while at the same time there are no poles in $\mathbb{C}_+$ or on the imaginary axis. This is an example where a larger delay value has a *stabilizing effect*. Considering all critical delays, it is easy to conclude that stability holds if and only if, for some non-negative integer ℓ, the delay τ satisfies

$$\frac{2\ell\pi}{\sqrt{\omega_0^2 - k}} < \tau < \frac{(2\ell + 1)\pi}{\sqrt{\omega_0^2 + k}}. \tag{3.16}$$

Let us next adopt the Rekasius transformation (3.11) for the same problem. In this case, the transformed characteristic equation becomes

$$(s^2 + \omega_0^2)(1 + Ts) - k(1 - Ts) = 0, \tag{3.17}$$

where $s = i\,\omega$. The above equation can be put in a Routh–Hurwitz table, parametric in T. From the table and using the ideas in [17, 18, 43], it is easy to see that both $T = T_1 = 0$ and $T = T_2 \to \infty$ are feasible solutions rendering roots at $s = i\,\omega$, respectively, with $\omega = \omega_{c,1}$ and $\omega = \omega_{c,2}$ as described above. The delay values corresponding to these crossings are then computed from (3.13) as

$$\tau_{c,1,\ell_1} = \frac{2}{\omega_{c,1}}(\tan^{-1}(\omega_{c,1}T_1) \mp \ell_1\pi) = \frac{2\ell_1\pi}{\omega_{c,1}}, \tag{3.18}$$

$$\tau_{c,2,\ell_2} = \frac{2}{\omega_{c,2}}(\tan^{-1}(\omega_{c,2}T_2) \mp \ell_2\pi) = \frac{2}{\omega_{c,2}}\left(\frac{\pi}{2} \mp \ell_2\pi\right). \tag{3.19}$$

Notice that the above equations consider both positive and negative solutions of delays. While this is mathematically feasible, only non-negative values are physically meaningful. Considering only those that are non-negative, some basic algebraic simplifications show that these results match with those presented above following (3.15). Moreover, in terms of crossing directions, we have that in the general case, all delay values associated with a crossing at ω cause the roots $s = i\,\omega$ to move only in one direction, either to the right-half or left-half complex plane [45]. This brings an organization to the behavior of root crossings, mainly creating a finite number of "root clusters" [17, 43] each designated by an imaginary axis crossing and its respective crossing direction, and where in each cluster one has infinitely many roots that correspond to that imaginary axis crossing for infinitely many, periodically spaced delay values. For instance, in the example at hand, all the delays $\tau_{c,1,\ell}$ cause stabilizing crossings associated with root clusters one, while all the delays $\tau_{c,2,\ell}$ yield destabilizing crossings associated with root clusters two. The effects of these delays on crossings eventually have the potential to lead to multiple stability intervals, as is the case in the example problem studied here; see (3.16).

3.2.4.3 A Two-Delay Example

As a third example, we consider a system with two distinct delays τ_1 and τ_2:

$$x''(t) + 2x'(t) + x(t) = -5x(t - \tau_1) - 4x'(t - \tau_2).$$

The characteristic equation is

$$f(s; \tau_1, \tau_2) := s^2 + 2s + 1 + 5e^{-\tau_1 s} + 4se^{-\tau_2 s} = 0. \tag{3.20}$$

We would like to find out for which delay pairs (τ_1, τ_2) the system is stable.

We demonstrate the solution by starting with the Rekasius transformation (3.11), but this time it is implemented for each exponential term independently [17, 43]. This transformation yields two parameters, $T_1 \in \mathbb{R}$ and $T_2 \in \mathbb{R}$ in the resulting transformed characteristic equation, which has the form

$$F := \sum_{k=0}^{4} \beta_k(T_1, T_2)s^k = 0, \tag{3.21}$$

where $\beta_k(T_1, T_2)$ are real coefficients. Differently from the single-delay case, we present an approach where frequency sweeping is leveraged. Letting $s = i\omega$ with ω being a sweep parameter, we incorporate the frequency-sweeping technique (see further ideas along these lines in [53]), and note that F can be zero only if its real and imaginary parts are zero for some T_1 and T_2,

$$\Re(F) := T_1(\omega^4 T_2 - 6\omega^2 + 4T_2\omega^2) - \omega^2 + 2T_2\omega^2 + 6,$$
$$\Im(F) := T_1(-\omega^3 + 2\omega^3 T_2 - 4\omega) - \omega^3 T_2 + 6\omega + 6T_2\omega.$$

The common roots of the above two equations can be computed using the *resultant theory* [54]. For this, one uses the coefficients of T_1 in $\Re(F)$ and $\Im(F)$ as the entries of Sylvester's matrix $\mathscr{S}$. The singularity of this matrix is a necessary condition for $\Re(F)$ and $\Im(F)$ to have a common root, which requires computing the determinant of $\mathscr{S}$. This determinant $\mathscr{R}_{T_1}(F) := \mathscr{R}_{T_1}(\Re(F), \Im(F))$, also known as the resultant of $\Re(F)$ and $\Im(F)$ by eliminating T_1, is given by

$$\mathscr{R}_{T_1}(F) := (-\omega^7 + 24\omega^3 - 2\omega^5)T_2^2 + (-16\omega^3 + 16\omega^5)T_2 - \omega^5 - 34\omega^3 + 24\omega.$$

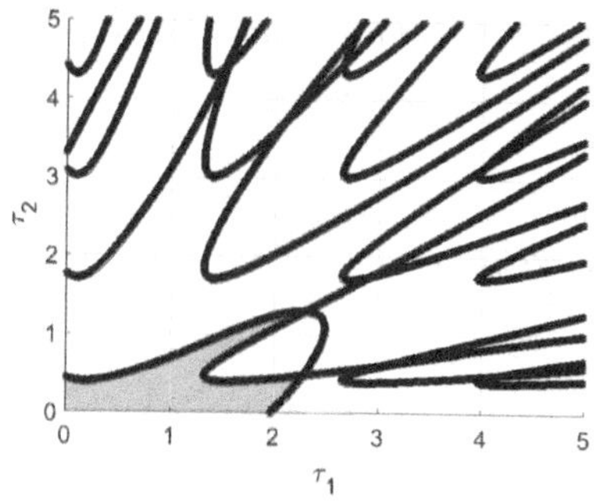

Fig. 3.1 Stability map of the system represented by (3.20). Stability region is shaded

Next, one sweeps the frequency ω and, for each given ω, performs the following: If there exists $T_2 \in \mathbb{R}$ such that $\mathscr{R}_{T_1} = 0$, then this T_2 is a candidate solution. A candidate T_1 is solved from $\Re(F) = 0$, and the candidates are checked for whether or not they satisfy $\Im(F) = 0$. If they do, then they are the common roots of $\Re(F)$ and $\Im(F)$; otherwise they are not. Once all common root pairs (T_1, T_2) corresponding to all ω are computed, one can then calculate all the delay values causing the crossing $s = i\omega$ using (3.13). Since feasible ω form continuous curves with respect to (T_1, T_2), the corresponding delay solutions will construct the well-known potential stability switching curves on the plane of τ_1-τ_2. Because the delay-free system, i.e., the system with $\tau_1 = \tau_2 = 0$ is stable, the region containing the origin of the delay-parameter space is guaranteed to be stable (gray shaded); see Fig. 3.1. The remaining regions should be analyzed following [16, 17, 20, 43, 55] and turn out to be unstable. Kernel and Offspring curves in *Cluster Treatment of Characteristic Roots* [43] can also be constructed from potential stability switching curves.[4]

Without question, guaranteeing the stability of dynamical systems is critical. The stability analysis above is valid in the linear regime, hence relates to how system states against small incoming perturbations would either damp out and vanish, or consistently grow to the point that the linearity assumption is no longer valid. In the context of traffic flow, these states correspond to the velocity and displacement of individual vehicles. In addition to these considerations, in interpreting the results it is important to note that the aforementioned perturbations are considered to be in impulsive form, as usually assumed in control theory. In other words, the analysis in this chapter pertains to impulse-response characteristics. However, perturbations can also be in cyclic form, and in a network of coupled systems, these perturbations would propagate. The relevant question then would be how the coupled system would behave under such cyclic perturbations. This calls for a frequency-response analysis, which will be presented in Chap. 6.

[4] See also [53, 56, 57] using tools in linear algebra and various forms of parameter scanning.

3.3 Interaction Between Multiple Delays and Arising Complexity

The manner in which time-delays interact with each other can unveil certain characteristics about the system's dynamics. For example, consider the following delay differential equation including two discrete delays:

$$x'(t) = A_0x(t) + A_1x(t - \tau_1) + A_2x(t - \tau_2), \tag{3.22}$$

under appropriate initial conditions, where $A_k \in \mathbb{R}^{n\times n}$, $k = 0, 1, 2$. In this case, the characteristic function in its general form reads

$$f(s; \tau_1, \tau_2) = P_0(s) + \sum_{\ell_1=1}^{n_1} P_{1,\ell_1}(s)e^{-\ell_1\tau_1 s} + \sum_{\ell_2=1}^{n_2} P_{2,\ell_2}(s)e^{-\ell_2\tau_2 s}$$
$$+ \sum_{\ell_1=1}^{n_1'}\sum_{\ell_2=1}^{n_2'} P_{3,\ell_1,\ell_2}(s, e^{-s\tau(\ell_1,\ell_2)}), \tag{3.23}$$

where P_0 is a nth-order polynomial, and $P_{i,\ell_i}(s)$, P_{3,ℓ_1,ℓ_2} are polynomials of lower-order. The function $\tau(\cdot,\cdot)$ defines appropriate combinations of τ_1 and τ_2 as

$$\tau : \mathbb{Z}_+ \times \mathbb{Z}_+ \mapsto \mathbb{R}_+, \qquad \tau(\ell_1, \ell_2) := \ell_1\tau_1 + \ell_2\tau_2. \tag{3.24}$$

Consider now a point $(\tau_{10}, \tau_{20}) \in \mathbb{R}_+^2$ in the delay-parameter space and denote by $\Gamma(\tau_{10}, \tau_{20}) \subset \mathbb{R}^2$ some neighborhood of it. We say that the system is *hyperbolic in* $\Gamma(\tau_{10}, \tau_{20})$ if the characteristic function f does not possess any imaginary zeros for all delays $(\tau_1, \tau_2) \in \Gamma(\tau_{10}, \tau_{20})$ [38].

It can be shown that, for any delay pair (τ_1, τ_2), the characteristic function $f(s; \tau_1, \tau_2)$ possesses a finite number of zeros in $\mathbb{C}_+$ [37]. Thus, if the system (3.22) is hyperbolic, the number of unstable zeros is *fixed* in the neighborhood $\Gamma(\tau_{10}, \tau_{20})$. If this number is zero the system (3.22) is stable, otherwise unstable.

In the sequel, we will briefly discuss two problems encountered when analyzing the stability of dynamical systems including multiple delays: weak/strong delay interactions and delay rays and delay interference. Weak delay interaction and delay interference concern cases when the hyperbolicity is valid for large delay values.

3.3.1 *Weak and Strong Delay Interactions*

In order to describe the way delays interact with each other, consider a simpler problem with the characteristic function

$$f(s; \tau_1, \tau_2) = P_0(s) + P_{1,1}(s)e^{-\tau_1 s} + P_{2,1}(s)e^{-\tau_2 s}, \tag{3.25}$$

where it is assumed that $f(0; \tau_1, \tau_2) \neq 0$; that is, f does not have a root at the origin. Without any lack of generality, assume that the corresponding system is not delay-independent hyperbolic. Let $\Delta(s, \tau_1) = P_0(s) + P_{1,1}(s)e^{-\tau_1 s}$ denote the first two terms in (3.25), and assume that $\Delta(s, \tau_1)$ is hyperbolic for $\tau_1 \in (\underline{\tau}_1, \overline{\tau}_1)$. Based on (3.25) and in view of hyperbolicity, it follows that for the system to be in transition from stability to instability, or vice versa, there must be at least one imaginary zero $s = i\omega$, in which case (3.25) by equating to zero and re-arranging becomes

$$1 + \frac{P_{2,1}(i\omega)}{\Delta(i\omega, \tau_1)} e^{-\omega\tau_2 i} = 0. \tag{3.26}$$

where, without loss of generality, $\omega > 0$ since $f(0; \tau_1, \tau_2) \neq 0$.

It is easy to see that for all $\tau_1 \in (\underline{\tau}_1, \overline{\tau}_1)$, hyperbolicity is preserved for all $\tau_2 \in \mathbb{R}_+$ if the inequality

$$\frac{|P_{2,1}(i\omega)|}{|\Delta(i\omega, \tau_1)|} < 1, \tag{3.27}$$

holds for all $\omega > 0$. The above property is established by observing that $\left(e^{-i\omega\tau_2}\right)|_{s=i\omega} = 1$ for all $\tau_1 \in \mathbb{R}_+$; root continuity holds [36]; and hyperbolicity is preserved if and only if there does not exist any imaginary axis crossings of (3.26).

Moreover, under the assumption of stability and hyperbolicity in $\tau_1 \in (\underline{\tau}_1, \overline{\tau}_1)$, we have by (3.27) that the system associated with (3.25) is stable *independently* of τ_2. In this case, we simply say that the *delay interaction* is *weak*. In other words, this situation corresponds to the existence of some unbounded stable regions in the delay-parameter space. If there exists at least one real ω satisfying the crossing condition (3.26), then the *delay interaction* is said to be *strong*.

In order to further illustrate the weak delay interactions, consider the scalar delay differential equation with $P_0(s) = s$, $P_{1,1}(s) = 1$, and $P_{2,1} = 1/2$. The corresponding characteristic function is

$$f(s; \tau_1, \tau_2) = s + e^{-\tau_1 s} + \frac{1}{2} e^{-\tau_2 s}. \tag{3.28}$$

Suppose that $\tau_1 \in (\pi/9, \pi/6)$. It is easy to see that $f(0; \tau_1, \tau_2) \neq 0$ and that the delay-free system is hyperbolic. Furthermore, it is stable since the only zero of (3.28) for zero delays is real and negative. However, the system is not delay-independent hyperbolic since $f(i\sqrt{3}/2; 4\pi/(3\sqrt{3}), 0) = 0$.

The frequency-sweeping condition (3.27) rewrites as

$$\frac{|P_{2,1}(i\omega)|}{|\Delta(i\omega, \tau_1)|} = \frac{1/2}{(i\omega + e^{-i\omega\tau_1})}. \tag{3.29}$$

Some simple but tedious algebraic manipulations show that

$$\frac{|P_{2,1}(i\omega)|}{|\Delta(i\omega, \tau_1)|} < \frac{1/2}{2/3} = \frac{3}{4} < 1, \quad \forall \tau_1 \in \left(\frac{\pi}{9}, \frac{\pi}{6}\right). \tag{3.30}$$

Thus, we will have a *weak delay interaction* and the system is stable for all delays in the "strip" $(\tau_1, \tau_2) \in (\pi/9, \pi/6) \times \mathbb{R}_+$.

One can also study the admissible frequency domains satisfying (3.27) for a given delay $\tau_1 \in \mathbb{R}_+$. The frequency domains where this condition holds can then provide insight into the geometry of the delay neighborhood $\tau_1 \in (\underline{\tau}_1, \overline{\tau}_1)$ for which some delay-independent stable directions may arise along τ_2. Such a geometry can be extracted using the argument conditions from (3.26). We leave this part of the analysis as an exercise for the reader.

Remark 3.3 It is important to note that weak delay interaction does not mean delay-independent hyperbolicity in the entire delay-parameter space, as proven based on the stability crossing curve characterizations [17, 43]. The existence of strong delay interactions may cause different types of stability results, as will be discussed in Chap. 4. We note that such a problem can be handled in the stability analysis following the techniques proposed in the literature (see, e.g., [1, 17, 43]). However, various degenerate cases need more care in their treatment and these discussions need to be put in the context of traffic flow dynamics; see Chaps. 5–6.

Remark 3.4 In the study of time-delay systems, delay sensitivity is as important as stability analysis. In this chapter, we have merely touched upon the basics of the stability analysis. Further work on delay sensitivity can be found in [14, 58–60].

3.3.2 *Delay Rays and Delay Interference*

Let us study (3.25) with $f(0; \tau_1, \tau_2) \neq 0$. Consider an arbitrary point $(\tau_{1,0}, \tau_{2,0}) \in \mathbb{R}_+^2 - \{(0, 0)\}$ in the delay-parameter space that is not located at the origin. Then the set

$$\mathscr{T}(\tau_{1,0}, \tau_{2,0}) := \left\{\alpha(\tau_{1,0}, \tau_{2,0}) \in \mathbb{R}_+^2 : \quad \alpha \in \mathbb{R}_+\right\} \tag{3.31}$$

defines a *ray* in the delay-parameter space $\mathbb{R}_+^2$ that includes the point $(\tau_{1,0}, \tau_{2,0})$. Assume that the delay parameters are restricted to "move" on the (delay) ray $\mathscr{T}(\tau_{1,0}, \tau_{2,0})$, that is, there exists some real $\alpha \geq 0$ such that $\tau_1 = \alpha\tau_{1,0}$, $\tau_2 = \alpha\tau_{2,0}$.

We say that the ray $\mathscr{T}(\tau_{1,0}, \tau_{2,0})$ is *stable* if the zero solution of the corresponding delay differential equation (3.22) is stable for all $(\tau_1, \tau_2) \in \mathscr{T}(\tau_{1,0}, \tau_{2,0})$. In other words, a stable delay ray corresponds to some particular *delay-independent stability* property in the delay-parameter space where the parameters are moving along the delay ray.

In this frame, one of the natural problems is to analyze the sensitivity of the delay-independent stability along a ray with respect to changes of the "direction" defined by $(\tau_{1,0}, \tau_{2,0})$. In particular, a stable delay ray is subject to *delay-interference phenomena* [29, 61] if the ray loses its stability property for arbitrarily small variations of the direction ray. Along these lines, see [62] where interference between two delay terms is attributed to promote stability. As discussed in [47], the stability of the delay rays in some delay-sector is equivalent to delay-independent stability with respect to all the delay parameters[5]; it appears that the stable rays are subject to delay-interference phenomena if the system is not delay-independent stable with respect to all delay parameters.

In order to better understand the delay-interference phenomenon, consider the scalar delay equation with $P_0(s) = s + 1$, $P_{1,1}(s) = 1$, $P_{2,1}(s) = 1/2$, borrowed from [47, 63], with the characteristic function

$$f(s; \tau_1, \tau_2) = s + 1 + e^{-\tau_1 s} + \tfrac{1}{2}e^{-\tau_2 s}. \tag{3.32}$$

Consider the delay ray $\mathscr{T}(1, 2)$ corresponding to some particular commensurate delay pair: $\tau_1 = \tau$, $\tau_2 = 2\tau$, i.e., the characteristic function $f(s; \tau, 2\tau)$. In this case, the stability of the delay ray $\mathscr{T}(1, 2)$ is equivalent to the delay-independent stability with respect to the delay τ. Since the delay-free system is stable, one needs to check if the system is delay-independent hyperbolic with respect to τ.

If $i\omega$, $\omega \neq 0$, is some root of $f = 0$ located on the imaginary axis, then the following conditions hold simultaneously for some delay $\tau \in \mathbb{R}_+$:

$$\omega = \sin(\omega\tau)(1 + \cos(\omega\tau)), \qquad \cos^2(\omega\tau) + \cos(\omega\tau) + \frac{1}{2} = 0. \tag{3.33}$$

Since the equation $z^2 + z + 1/2 = 0$ does not have real solutions in $[-1, 1]$, no real ω exists satisfying (3.33), and, in conclusion, the delay ray $\mathscr{T}(1, 2)$ is stable.

Consider now a perturbation of the delay ray $\mathscr{T}(1, 2 + \varepsilon_n)$, for $n \in \mathbb{N}$, with $\varepsilon_n = \frac{1}{2(2n+1)}$. Clearly, $\varepsilon_n \to 0$ as $n \to \infty$. Furthermore,

$$f\left(i\frac{1}{2};\ 2(2n + 1)\pi,\ 2(2n + 1)(2 + \varepsilon_n)\pi)\right) = 0.$$

As a consequence, the delay ray $\mathscr{T}(1, 2 + \varepsilon_n)$ is not stable. Thus, the stable delay ray $\mathscr{T}(1, 2)$ is subject to delay-interference phenomenon. For a deeper analysis, the reader is referred to [29, 61].

3.4 Notes and Further Reading

Observations on the above developments, specifically from the viewpoint of multiple delays, can be summarized as follows. Although a single delay may be destabilizing, multiple delays with weak interference may offer stability independent of one of

[5] This is sometimes called *absolute stability* in the delay-parameter space.

the delays. Multiple delays may offer larger stability regions in the delay-parameter space compared to single delays.

Next, strong interaction/interference between delays may be advantageous and expand the stability boundaries of traffic flow dynamics.

Finally, multiple delays can be regarded as multiple parameters affecting the dynamics. Under certain settings and different combinations, these parameters may provide enhancement in certain metrics related to stability, performance, robustness, delay-independent stability, and disturbance rejection capabilities.

We note that the developments in traffic flow dynamics with time-delays have progressed partly independently of the advancements in time-delay systems in control theory. We believe that the above discussions are useful from this perspective, making the connections more transparent while bringing tools from control theory over to traffic flow dynamics.

These connections have been recognized by a number of publications, e.g., [64–69]. We hope that this brief serves to expand this very same philosophy.

For further reading on stability robustness, stabilization, delay-interference phenomenon, and delay-independent stability, the readers are referred to [14, 16, 17, 20, 31, 32, 43, 53, 55, 60, 61, 70, 71]. For an overview on the stability and stabilization of time-delay systems in the discrete delay case, the readers are referred to [1].

References

1. R. Sipahi, S.-I. Niculescu, C.T. Abdallah, W. Michiels, K. Gu, Stability and stabilization of systems with time delay, limitations and opportunities. IEEE Control Syst. Mag. **31**(1), 38–65 (2011)
2. A. Callender, D.R. Hartree, A. Porter, Time lag in a control system. Philos. Trans. R. Soc. Lond. Ser. A, Math. Phys. Sci. **235**, 415–444 (1936)
3. T. Erneux, *Applied Delay Differential Equations* (Springer, New York, 2009)
4. J.D. Sterman, *Business Dynamics: Systems Thinking and Modeling for a Complex World* (McGraw-Hill, Boston, 2000)
5. N. MacDonald, *Biological Delay Systems: Linear Stability Theory* (Cambridge University Press, Cambridge, 1989)
6. Y. Kuang, *Delay Differential Equations with Applications in Population Dynamics* (Academic Press, Boston, 1993)
7. K. Masani, A.H. Vette, N. Kawashima, M.R. Popovic, Neuromusculoskeletal torque-generation process has a large destabilizing effect on the control mechanism of quiet standing. J. Neurophysiol. **100**, 1465–1475 (2008)
8. N. Olgac, R. Sipahi, A.F. Ergenc, Delay scheduling: a new concept for stabilization in multiple delay systems. J. Vib. Control **11**, 1159–1172 (2005)
9. G. Stépán, T. Insperger, Stability of time-periodic and delayed systems—a route to act-and-wait control. Annu. Rev. Control **30**, 159–168 (2006)
10. C. Abdallah, P. Dorato, J. Benites-Read, R. Byrne, Delayed positive feedback can stabilize oscillatory systems, *American Control Conference* (San Francisco, CA, 1993), pp. 3106–3107
11. N. Olgac, B. T. Holm-Hansen, A novel active vibration absorption technique: delayed resonator. J. Sound Vib. **176**, 93–104 (1994)
12. A. Ramirez, S. Mondie, R. Garrido, R. Sipahi, An analytical approach to tuning of delay-based controllers for LTI-SISO systems. SIAM Journal on Control Optim. **55**(1), 397–412 (2017)

13. P. Zítek, J. Fišer, T. Vyhlídal, Dimensional analysis approach to dominant three-pole placement in delayed PID control loops. J. Process Control **23**(8), 1063–1074 (2013)
14. S.-I. Niculescu, W. Michiels, Stabilizing a chain of integrators using multiple delays. IEEE Trans. Autom. Control **49**(5), 802–807 (2004)
15. O. Toker, H. Özbay, Complexity issues in robust stability of linear delay-differential systems. Math. Control Signals Syst. **9**, 386–400 (1996)
16. K. Gu, S.-I. Niculescu, J. Chen, On stability crossing curves for general systems with two delays. J. Math. Anal. Appl. **311**, 231–253 (2005)
17. R. Sipahi, N. Olgac, Complete stability robustness of third-order LTI multiple time-delay systems. Automatica **41**, 1413–1422 (2005)
18. R. Sipahi, N. Olgac, Stability robustness of retarded LTI systems with single delay and exhaustive determination of their imaginary spectra. SIAM J. Control. Optim. **45**, 1680–1696 (2006)
19. K.L. Cooke, P. Van Den Driessche, On zeros of some transcendental equations. Funkcialaj Ekvacioj **29**, 77–90 (1986)
20. G. Stépán, *Retarded Dynamical Systems: Stability and Characteristic Function* (Longman Scientific & Technical, Wiley, US, New York, 1989)
21. A. Thowsen, The Routh-Hurwitz method for stability determination of linear differential-difference systems. Int. J. Control **33**, 991–995 (1981)
22. G. Orosz, Connected cruise control: modelling, delay effects, and nonlinear behaviour. Veh. Syst. Dyn. **54**(8), 1147–1176 (2016)
23. L. Zhang, G. Orosz, Motif-based design for connected vehicle systems in presence of heterogeneous connectivity structures and time delays. IEEE Trans. Intell. Transp. Syst. **17**(6), 1638–1651 (2016)
24. L. Zhang, J. Sun, G. Orosz, Hierarchical design of connected cruise control in the presence of information delays and uncertain vehicle dynamics. IEEE Trans. Control Syst. Technol. **26**(1), 139–150 (2017)
25. G. Orosz, E. Wilson, G. Stépán (eds.), Traffic jams—dynamics and control. Philos. Trans. R. Soc. A **368**(1928) (2010)
26. F. Kappel, *Linear Autonomous Functional Differential Equations*, Course notes
27. D. Breda, S. Maset, R. Vermiglio, *Stability of Linear Delay Differential Equations: A Numerical Approach with MATLAB*, SpringerBriefs in Control, Automation and Robotics (2014)
28. J.K. Hale, S.M. Verduyn Lunel, *Introduction to Functional Differential Equations* (Springer, New York, 1993)
29. W. Michiels, S.-I. Niculescu, *Stability and Stabilization of Time-Delay Systems* An Eigenvalue Based Approach. (SIAM, Philadelphia, 2007)
30. R. Bellman, K.L. Cooke, *Differential-Difference Equations* (Academic Press, New York, 1963)
31. K. Gu, V.L. Kharitonov, J. Chen, *Stability of Time-Delay Systems* (Birkhäuser Boston, Cambridge, MA, 2003)
32. R. Sipahi, *Mastering Frequency Domain Techniques for the Stability Analysis of LTI Time Delay Systems*, vol. 20 (SIAM, Philadelphia, 2019)
33. R. Datko, A procedure for determination of the exponential stability of certain differential-difference equations. Q. Appl. Math. **36**, 279–292 (1978)
34. K. Knopp, *Theory of Functions* (Dover, Mineola, NY, 1996)
35. T. Kato, *Perturbation Theory of Linear Operators*, 2nd edn. (Springer, Berlin, 1966)
36. L.E. El'sgol'ts, S.B. Norkin, *Introduction to the Theory and Applications of Differential Equations with Deviating Arguments* (Academic Press, New York, 1973)
37. J.K. Hale, E.F. Infante, F.-S. Peter Tsen, Stability in linear delay equations. J. Math. Anal. Appl. **105**, 533–555 (1985)
38. S.-I. Niculescu, *Delay Effects on Stability: A Robust Control Approach* (Springer, Heidelberg, 2001)
39. C.S. Hsu, Application of the tau-decomposition method to dynamical systems subjected to retarded follower forces. ASME J. Appl. Mech. **37**, 259–266 (1970)
40. C. Foley, M.C. Mackey, Dynamic hematological disease: a review. J. Math. Biol. **58**, 285–322 (2009)

41. J. Neimark, D-subdivisions and spaces of quasi-polynomials. Prikl. Mat. Mekh., **13**, 349–380 (1949)
42. M.S. Lee, C.S. Hsu, On the τ-decomposition method of stability analysis for retarded dynamical systems. SIAM J. Control **7**, 242–259 (1969)
43. R. Sipahi, *Cluster Treatment of Characteristic Roots, CTCR, A Unique Methodology for the Complete Stability Robustness Analysis of Linear Time Invariant Multiple Time Delay Systems Against Delay Uncertainties*, Ph.D. Dissertation, University of Connecticut, Mechanical Engineering Department (2005)
44. J.K. Hale, W.Z. Huang, Global geometry of the stable regions for two delay differential equations. J. Math. Anal. Appl. **178**, 344–362 (1993)
45. N. Olgac, R. Sipahi, An exact method for the stability analysis of time delayed Linear Time-Invariant (LTI) systems. IEEE Trans. Autom. Control **47**, 793–797 (2002)
46. Z.V. Rekasius, A stability test for systems with delays, *Joint Automatic Control Conference*, Paper No. TP9-A (1980)
47. J. Louisell, Absolute stability in linear delay-differential systems: Ill-possedness and robustness. IEEE Trans. Autom. Control **40**, 1288–1291 (1995)
48. J. Chen, H.A. Latchman, Frequency sweeping tests for stability independent of delay. IEEE Trans. Autom. Control **40**, 1640–1645 (1995)
49. J. Louisell, A matrix method for determining the imaginary axis eigenvalues of a delay system. IEEE Trans. Autom. Control **46**, 2008–2012 (2001)
50. K.E. Walton, J.E. Marshall, Direct method for TDS stability analysis. IEE Proc. D **134**, 101–107 (1987)
51. K.L. Cooke, Z. Grossman, Discrete delay, distributed delay and stability switches. J. Math. Anal. Appl. **86**(2), 592–627 (1982)
52. J. Chen, P. Fu, S.-I. Niculescu, When will zeros of time-delay systems cross imaginary axis?, *European Control Conference* (Kos, Greece, 2007), pp. 5631–5638
53. I.I. Delice, R. Sipahi, Delay-independent stability test for systems with multiple time delays. IEEE Trans. Autom. Control **57**(4), 963–972 (2011)
54. I.M. Gelfand, M.M. Kapranov, A.V. Zelevinsky, *Discriminants, Resultants, and Multidimensional Determinants* (Birkhäuser, Boston, MA, 1994)
55. T. Insperger, G. Stépán, *Semi-Discretization for Time-Delay Systems Stability and Engineering Applications* Applied Mathematical Sciences. (Springer, Berlin, 2011)
56. H. Fazelinia, R. Sipahi, N. Olgac, Stability analysis of multiple time delayed systems using 'Building Block' concept. IEEE Trans. Autom. Control **52**, 799–810 (2007)
57. H.Y. Hu, Z.H. Wang, *Dynamics of Controlled Mechanical Systems with Delayed Feedback* (Springer, Heidelberg, Germany, 2002)
58. W. Michiels, S.-I. Niculescu, L. Moreau, Using delays and time-varying gains to improve the output feedback stabilizability of linear systems: a comparison. IMA J. Math. Control Inf. **21**(4), 393–418 (2004)
59. W. Michiels, S.-I. Niculescu, On the delay sensitivity of Smith predictors. Int. J. Syst. Sci. **34**(8–9), 543–551 (2003)
60. W. Michiels, *Stability and stabilization of time-delay systems*, Ph.D. Thesis, Katholieke Universiteit Leuven, Belgium (2002)
61. W. Michiels, S.-I. Niculescu, Characterization of delay-independent stability and delay-interference phenomena. SIAM J. Control Optim. **45**, 2138–2155 (2007)
62. N. MacDonald, An interference effect of independent delays. IEE Proc. D **134**, 38–42 (1987)
63. R. Datko, Remarks concerning the asymptotic stability and stabilization of linear delay differential equations. J. Math. Anal. Appl. **111**, 571–584 (1985)
64. G. Orosz, G. Stépán, Hopf bifurcation calculations in delayed systems with translational symmetry. J. Nonlinear Sci. **14**(6), 505–528 (2004)
65. G. Orosz, R.E. Wilson, B. Krauskopf, Global bifurcation investigation of an optimal velocity traffic model with driver reaction time. Phys. Rev. E **70**, 026207 (2004)
66. P. Seiler, A. Pant, K. Hedrick, Disturbance propagation in vehicle strings. IEEE Trans. Autom. Control **49**(10), 1835–1842 (2004)

67. D. Swaroop, J.K. Hedrick, S.B. Choi, Direct adaptive longitudinal control of vehicle platoons. IEEE Trans. Veh. Technol. **50**(1), 150–161 (2001)
68. J. Zhang, P. Ioannou, Control of heavy-duty trucks: environmental and fuel economy considerations (University of Southern California, Los Angeles TO4203, 2004)
69. A. Bose, P.A. Ioannou, Analysis of traffic flow with mixed manual and semi automated vehicles. IEEE Trans. Intell. Transp. Syst. **4**(4), 173–188 (2003)
70. R. Sipahi, I.I. Delice, Extraction of 3D stability switching hypersurfaces of a time delay system with multiple fixed delays. Automatica **45**, 1449–1454 (2009)
71. P.M. Nia, R. Sipahi, Controller design for delay-independent stability of linear time-invariant vibration systems with multiple delays. J. Sound Vib. **332**(14), 3589–3604 (2013)

Chapter 4
Linear Stability of Traffic Flow Models with Discrete Delays

In this chapter, we investigate the linear stability of a number of car-following models by combining the material of Chaps. 2–3. The discussion here pertains to discrete-time delays, and also expands on the existing models by considering the effects of multiple delays that may arise from different sensing capabilities of the drivers. In particular, we study headway and relative-velocity sensing delays. We also study how following multiple vehicles ahead of traffic affects the model characteristics and stability.

From a multiple-delay point of view, our aim is to emphasize and study the robustness of the system under uncertainty in delays, mainly using stability maps. Although the main discussion assumes that drivers adopt a constant-headway strategy, further remarks are included on constant time-headway driving strategy where a driver's goal is to leave sufficient headway proportional to the speed of the driver's vehicle. Furthermore, the stability results established in this chapter prepare for the slinky analysis in Chap. 6.

To be consistent with the literature on car-following dynamics, we use the notation of $\dot{x}(t)$ instead of $x'(t)$ to denote the time derivative of a variable $x(t)$. The presentation here focuses on the stability analysis of the linearized dynamics of the corresponding models. For linearization of various models, readers are referred to the discussions in Chap. 2. To this end, depending on the type of model considered, a car-following dynamics has a constant velocity equilibrium point, and the linear stability analysis is concerned with understanding the dynamics of deviations from this point. While we use $x(t)$ and $\dot{x}(t) = v(t)$ to denote respectively the perturbations in position and velocity, these variables may also be used in nonlinear dynamics to maintain consistency with the literature. However, the use of the variable is clear from the context, otherwise this is clarified within the presentation. In the rest of the book, we will assume that vehicles are sufficiently apart from each other and investigate collision-free driving conditions around the equilibrium.

R. Sipahi et al., *Deterministic Car-Following Traffic Models*,
SpringerBriefs in Control, Automation and Robotics,
https://doi.org/10.1007/978-3-031-58164-9_4

4.1 Multiple-Vehicle-Following Strategy with a Single Time Delay

4.1.1 Mathematical Model

Our starting point is the car-following model from Chap. 2:

$$\dot{v}_k(t) = \alpha_k(v_{k+1}(t-\tau) - v_k(t-\tau)). \tag{4.1}$$

Differently, we take into account that the drivers may observe and follow multiple vehicles ahead of traffic [1]. Based on (4.1), the following model for multiple vehicle following strategy can be proposed (see also [2]):

$$\dot{v}_k(t) = \sum_{j=1}^{N} \alpha_{k,j}(v_{k+j}(t-\tau) - v_k(t-\tau)), \tag{4.2}$$

where N is the number of vehicles being followed, $\alpha_{k,j}$ are the drivers' aggressiveness constants depending on which vehicle they are following, and the time delay $\tau \in \mathbb{R}^+$ is assumed to be identical for all the drivers.

In (4.2), the delay τ can be interpreted as an "input delay"; that is, the velocity discrepancy $v_{k+j}(t) - v_k(t)$ between the two vehicles is sensed first, but is then subject to a delay imposed by the driver's physiology (see Chap. 2 for further details). In the following development, we assume for definiteness that six vehicles ($n = 6$) are arranged around a ring and vehicles are operated by identical drivers.[1] The constants $\alpha_{k,j}$ are taken as $\alpha_{k,1} = \gamma$ for all k, and $\alpha_{k,j} = \beta$ for all k and $j > 1$. Here, the rationale is that speed adjustments based on the immediately preceding vehicle do not need to be equally weighted compared to the adjustments based on the further upstream vehicles. The aim is then to analytically study the stability properties of the arising dynamics with respect to the single delay parameter τ as the number of followed vehicles N is increased up to five, $N = 1, \ldots, 5$.

4.1.2 Stability Analysis Results

Considering the circular path in Fig. 1.1 with $n = 6$ vehicles and given N, the characteristic equation of the dynamics can easily be obtained. On the other hand, this equation is of sixth order in the Laplace variable s and carries commensurate delays,

[1] The subscript notation in (4.2), and throughout this book, is presented conveniently for the linear configuration of vehicles. When adapting the model for the ring configuration, one needs to pay attention to the circular pattern of vehicles and thus properly label the subscripts modulo n.

i.e., exponential terms involving exponents $\tau, 2\tau, 3\tau, \ldots$. In problems where the total number of vehicles n is large, the analysis of the characteristic equation can be computationally restrictive. A remedy arises when one observes that the system dynamics at hand has a particular structure, as demonstrated next. Specifically, (4.2) can be written in vector form as

$$\frac{dv}{dt} = Av(t - \tau), \tag{4.3}$$

where A is the configuration matrix, which is equal to the negative Laplacian matrix of the corresponding vehicle network. In case A is diagonalizable, the characteristic function can be written in the factored form

$$\det[sI - Ae^{-s\tau}] = \prod_{\ell=1}^{n}(s - \lambda_\ell(A)e^{-s\tau}). \tag{4.4}$$

Thus, one can conveniently study each factor on the right-hand side separately. Since the Laplacian always has a zero eigenvalue, $\lambda_1(A) = 0$, then $s = 0$ is always a characteristic root, which can be ignored in the stability analysis since it simply corresponds to the rigid body dynamics of the ensemble of vehicles as a whole [3–6]. Moreover, with $\gamma > 0$ and $\beta > 0$, and given that the vehicle network is connected, we have that $\lambda_\ell(A) \in \mathbb{C}_- := \{s \in \mathbb{C} : \Re(s) < 0\}$, $\ell = 2, \ldots, n$ [7]. That is, in polar form $\lambda_\ell = \rho_\ell e^{i\phi_\ell}$, where $\phi_\ell \in (\pi/2, 3\pi/2)$ and $\rho_\ell > 0$. One can then formulate the smallest positive delay causing an imaginary axis crossing in of each subsystem as

$$\tau_\ell^* = \frac{\phi_\ell - \pi/2}{\rho_\ell} > 0, \quad \phi_\ell(\pi/2, \pi] \quad \ell = 2\ldots, n, \tag{4.5}$$

where the range of ϕ_ℓ is selected such that one considers only the eigenvalues on the second quadrant of the complex plane and on the negative real axis. This is simply because the complex eigenvalues are symmetric with respect to the Real axis. Consequently, the delay margin of the entire system is the smallest of delay margins of all subsystems,

$$\tau^* = \min_{\ell=2,\ldots,n} \tau_\ell^*,$$

and the eigenvalue that corresponds to this delay margin is called the *responsible eigenvalue* [3–6]. When A has only real eigenvalues, one recovers the well-known result of $\tau^* = \pi/(2 \max_\ell |\lambda_\ell|)$ [8].

In light of the above, the *delay margin* of the car-following dynamics can easily be studied as a function of the number of vehicles being followed; see Fig. 4.1. We find out that, under the problem assumptions made above, following more than one vehicle upstream of traffic causes *smaller* delay margins. It is therefore preferable that the drivers follow only the vehicle immediately in front of them, since this selection offers a larger stability margin for the entire dynamics.

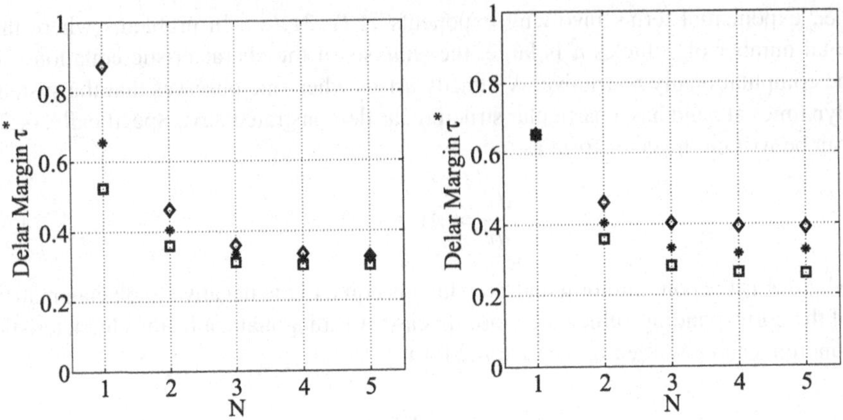

Fig. 4.1 Delay margin of the dynamics (4.2) as a function of N, the number of vehicles each vehicle follows. Driver's aggressiveness constants are γ and β. Left panel: $\beta = 0.8$ is fixed, and $\gamma = 0.6$ (diamond); $\gamma = 0.8$ (star); $\gamma = 1.0$ (square). Right panel: $\gamma = 0.8$ is fixed, $\beta = 0.6$ (diamond); $\beta = 0.8$ (star); $\beta = 1.0$ (square)

In Fig. 4.1, we explore how the delay margin is affected for different numerical settings. We see that the delay margin can be enhanced by reducing γ (left panel), that is, by reducing the driver's aggressiveness in speed corrections with respect to the vehicle immediately in front of it. Next, we fix $\gamma = 0.8$ and explore how β affects the results; see the right panel where star markers represent the same conditions as those in the left panel. Here, we notice similar trends where reduced aggressiveness in β increases the delay margin of the system. Larger N is overall unfavorable since it causes the delay margin to shrink.[2]

4.2 Following Two Vehicles with Two Discrete Delays

4.2.1 Mathematical Model

We now study the effects of multiple delays on an extended version of the dynamics in (4.2). Here, all the drivers follow the two vehicles ahead of them; however sensing the nearer vehicle occurs with a shorter delay. Thus there exist two independent time delays τ_1 and τ_2. The main motivation for this assumption is that the drivers will pay more attention to the vehicle immediately in front of them than to a vehicle further away. In this section, we will focus on velocity feedback only, in accordance with

[2] It is worth noting that as the number of vehicles N followed increases, the system is subject to higher gains. Dynamical systems often have weaker tolerance against delays in the presence of high gains. One can scale the gains by $1/N$ to investigate how the delay margin is affected. This is left as an exercise for the reader.

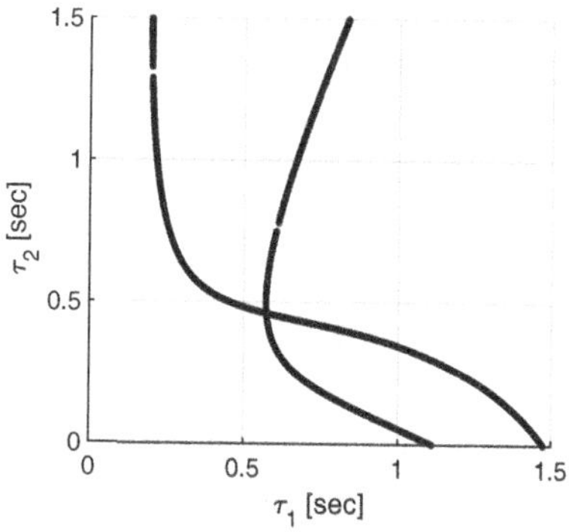

Fig. 4.2 Stability switching curves under velocity feedback corresponding to (4.6) with $\gamma = 1.2$ and $\beta = 0.75$. Stability region is connected to the origin of the delay plane

several models surveyed in Chap. 2. As an illustrative example, we consider $n = 3$ vehicles are arranged around a circular path:

Drivers follow two vehicles ahead of them by performing *only velocity feedback*. The linear dynamics read

$$\ddot{x}_k(t) = \sum_{j=1}^{2} \alpha_{k,j}(\dot{x}_{k+j}(t - \tau_j) - \dot{x}_k(t - \tau_j)). \tag{4.6}$$

The objective here is to understand how the driving strategy in (4.6) influences linear stability in the presence of two discrete delays, and, specifically, how the stability regions in the parameter space of delays are formed. For demonstration purposes, we choose $\alpha_{k,1} = \gamma = 1.2$ and $\alpha_{k,2} = \beta = 0.75$.

4.2.2 *Stability Analysis Results*

Referring to Chap. 3 for details on the stability analysis, we highlight the main results on the driving strategy (4.6). The dynamics in (4.6) is stable in the region connected to the origin of the τ_1–τ_2 plane, Fig. 4.2. Here we reveal that the dynamics has delay independent stability (or, in other words an *unbounded* stability region) along the τ_2 axis when $\tau_1 < 0.196$ s. This means that if the drivers could act no later than 0.196 s of delay in their decisions associated with velocity feedback, then the dynamics would remain stable regardless of how late they act in their decisions associated with vehicles further ahead. However, this delay amount is smaller than typical human reaction delays; hence, it may not be physically possible for such a stabilization mechanism to actually take effect.

Moreover, based on the assumption that τ_1 is smaller than τ_2, one can divide the stability region into two equal parts by using a 45° line emanating from the origin, and focus on the region where $\tau_2 > \tau_1$. If $0.196 < \tau_1 < 0.48$ sec, stability is no longer independent of τ_2 and decreasing τ_1 allows larger τ_2 values for which stability is maintained. Clearly, this points out a trade-off where the delays cannot be made larger independently and at the same time retaining stability.

4.3 Robustness in the Presence of Delay Mismatch

We now look into a slightly different problem. In the previous subsection, the delay terms were different when following different vehicles. Here, however, we investigate a delay mismatch in a particular action of the driver, e.g., different delays in sensing own velocity versus the preceding vehicle's velocity. We first start with a conceptual circular path traffic scenario, and then investigate *weak* and *strong interactions between time delays* [2].

4.3.1 *Weak and Strong Delay Interactions*

We start by focusing on the model in [9] with a single delay, and extend the result by studying how a delay mismatch in the model could affect stability. This leads to:

Two-delay car-following model with delay mismatch

$$\dot{v}_k(t) = \alpha_k(v_{k+1}(t-\tau_2) - v_k(t-\tau_1)), \tag{4.7}$$

where $\tau_1 - \tau_2$ is the delay mismatch.

For definiteness, we study a scenario with three vehicles, which yields system equations in vector form

$$\dot{v}(t) = \begin{pmatrix} -\alpha_1 & 0 & 0 \\ 0 & -\alpha_2 & 0 \\ 0 & 0 & -\alpha_3 \end{pmatrix} v(t-\tau_1) + \begin{pmatrix} 0 & 0 & \alpha_1 \\ \alpha_2 & 0 & 0 \\ 0 & \alpha_3 & 0 \end{pmatrix} v(t-\tau_2), \tag{4.8}$$

where $v(t) = [v_1(t), v_2(t), v_3(t)]^\top$. We pick the constants[3] $\alpha_1 = 1.6$, $\alpha_2 = 1.9$, and $\alpha_3 = 2.2$, consistent with [10], which yields the system characteristic function

$$f = s^3 + 5.7e^{-\tau_1 s}s^2 + 10.74e^{-2\tau_1 s}s + 6.688(e^{-3\tau_1 s} - e^{-3\tau_2 s}). \tag{4.9}$$

[3] For a different combination of parameters, see [2].

Note that the non-delayed (delay-free) dynamics, where $\tau_1 = \tau_2 = 0$, as expected from the "consensus" nature of the model, is marginally stable, with poles at $s_1 = 0$ and $s_{2,3} = -2.85 \pm 1.618i$. Moreover, for some interval $\tau_1 \in (\underline{\tau}_1, \overline{\tau}_1)$, one can observe a *weak delay interaction* appearing along the τ_2 direction, where delay-independent stability feature appears. This is related to (3.27): it is easy to show that for sufficiently large ω,

$$\frac{6.688}{|(i\omega)^3 + 5.7e^{-\tau_1 \omega i}(i\omega)^2 + 10.74e^{-2\tau_1 \omega i} i\omega|} < \frac{1}{2} \tag{4.10}$$

holds regardless of τ_2. With computations, we can also show that the above inequality holds for $\omega > 2.5$. However, for sufficiently small ω, we have $\tau_1 > \overline{\tau}_1$ and hence this feature will disappear and *strong delay interactions* will arise.

To further understand and visualize the weak and strong delay interactions, we look at the situation in the parameter plane of delays (see the details in Chap. 3). As mentioned above, a zero pole $s_1 = 0$ is *invariant* no matter how time delays are selected, since $e^{-\tau s} = 1$ at $s = 0$ for all $\tau \geq 0$, corresponding to the rigid body dynamics of the ensemble of vehicles, when they have identical velocities. Nevertheless, there is the possibility of having another zero pole for some non-zero delays [11]. To investigate whether or not this issue arises here, we inspect the characteristic function for the double root condition at $s = 0$. For this particular choice of parameters, the condition reads

$$\left.\frac{df}{ds}\right|_{s=0} = 0.535 - \tau_1 + \tau_2 = 0, \tag{4.11}$$

which defines a line on the delay plane (Fig. 4.3), with a slope of 45°. As the definition implies, the necessary condition for the dynamics to have a double root at $s = 0$ is that the delays lie on the line (4.11). As a matter of fact, since the characteristic function (4.9) is already satisfied at $s = 0$ due to the root invariance with respect to any delay value, (4.11) is indeed necessary and sufficient for the existence of a double root at $s = 0$. Further analysis should be pursued considering two key questions (see [11] for further details):

(a) *In which direction does the second pole at the origin cross the origin?* This can be studied by either developing a sensitivity analysis at $s = 0$ for the given delay values defined in Eq. (4.11), or observing that one side of the line borders the stability region connected with the origin of the delay plane. That is, the right-hand side of the line equation bordering the stability region must have a positive real pole, which crosses over at $s = 0$ as the delay parameter varies across this line.
(b) *Could there be triple, quadruple, etc., roots exhibiting a similar behavior?* For this, one needs to perform a similar analysis as in (4.11) and check if higher order derivatives simultaneously vanish. In the case of triple roots, this would require $(d^2 f/ds^2)|_{s=0} = 0$ to also hold for some delay values satisfying the double root

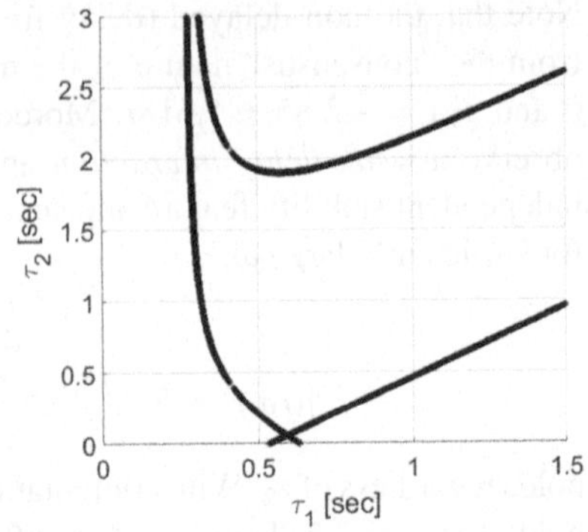

Fig. 4.3 Stability boundaries of the dynamics (4.8) where stability region is connected to the origin of the delay plane. Notice the line Eq. (4.11) representing the delay combinations at which the system has double zero poles

condition (4.11). Using simple algebraic manipulations, it is easy to show for the particular problem at hand that the presence of triple or of higher order roots at the origin of the complex plane is not possible.

Next, we extract the stability map of the given dynamics in Fig. 4.3 using its characteristic function and the tools in Chap. 3. Note on this figure that the curves, *except* the double root line equation Eq. (4.11), represent the boundaries where the characteristic function (4.9) has at least one pair of zeros $s = \pm i\omega$ on the imaginary axis. Moreover, following the guidelines in Chap. 3, one can identify which regions on the delay plane the dynamics is stable. For example, since the system is stable for zero delays "around" its rigid body dynamics (which is associated with the zero pole), it remains stable around its rigid body dynamics as long as the delays are located in the region connected to the origin.

4.3.2 *Strong Interaction Between Delays and Trade-Offs*

We now consider the model proposed by Bando and colleagues [10] for a circular path of length L and for $n = 3$ vehicles. The major difference here is that we use two independent delays, as reasoned above, as opposed to a single delay in the cited reference. This revises the nonlinear model as

$$\ddot{q}_k(t) = \alpha_k(V(q_{k+1}(t-\tau_1) - q_k(t-\tau_1)) - \dot{q}_k(t-\tau_2)), \tag{4.12}$$

where the nonlinear function $V(\Delta x)$ can be selected similarly as in Chap. 1, and $q_k(t)$ represents the position of the vehicle k measured from a fixed reference frame.

The objective is to study the linear stability of the perturbations $x(t)$ around the equilibrium of the nonlinear dynamics (4.12), in the parameter space of the delays τ_1 and τ_2. For this, we first linearize (4.12) around its equilibrium. The equilibrium conditions are determined when the vehicles reach a constant speed, i.e., zero acceleration, with equilibrium headway given by $\Delta x_{eq} = L/n$. Note that time delay has no effect on the equilibrium value since velocity and headway are time-invariant at the equilibrium. Hence, by setting (4.12) to zero one can extract the equilibrium point for the constant velocity of the vehicles.

Next, we introduce *small perturbations* $x(t)$ and $v(t)$ around the equilibrium headway and velocity, and obtain the following linear set of delay differential equations representing their dynamics

$$\dot{v}_k(t) = \alpha_k(\alpha_0(x_{k+1}(t-\tau_1) - x_k(t-\tau_1)) - v_k(t-\tau_2)), \tag{4.13}$$

$$\dot{x}_k(t) = v_k(t), \tag{4.14}$$

where $\alpha_0 = (dV/d\Delta x)|_{\Delta x = \Delta x_{eq}}$ is the slope of the optimal velocity function at the equilibrium headway. Considering a circular path, we can express (4.13)–(4.14) in state space form for three vehicles as follows:

$$\begin{pmatrix} \dot{x}(t) \\ \dot{v}(t) \end{pmatrix} = \begin{pmatrix} 0_{3\times3} & I_{3\times3} \\ 0_{3\times3} & 0_{3\times3} \end{pmatrix} \begin{pmatrix} x(t) \\ v(t) \end{pmatrix}$$
$$+ \begin{pmatrix} 0_{3\times3} & 0_{3\times3} \\ D_{3\times3} & 0_{3\times3} \end{pmatrix} \begin{pmatrix} x(t-\tau_1) \\ v(t-\tau_1) \end{pmatrix} + \begin{pmatrix} 0_{3\times3} & 0_{3\times3} \\ 0_{3\times3} & C_{3\times3} \end{pmatrix} \begin{pmatrix} x(t-\tau_2) \\ v(t-\tau_2) \end{pmatrix},$$

where $0_{3\times3}$ is the 3×3 zero matrix, $I_{3\times3}$ is the 3×3 identity matrix, and the matrices C and D are defined as

$$C = \begin{pmatrix} -\alpha_1 & 0 & 0 \\ 0 & -\alpha_2 & 0 \\ 0 & 0 & -\alpha_3 \end{pmatrix}, \qquad D = \alpha_0 \begin{pmatrix} -\alpha_1 & \alpha_1 & 0 \\ 0 & -\alpha_2 & \alpha_2 \\ \alpha_3 & 0 & -\alpha_3 \end{pmatrix}. \tag{4.15}$$

Based on the above settings and for $\alpha_0 = 0.7$, $\alpha_k = 1.8$, $k = 1, 2, 3$, the characteristic function of the dynamics can be obtained as

$$f = \frac{1}{125} s(5\,s + 9e^{-\tau_2 s}) \times (25s^4 + 90e^{-\tau_2 s}s^3 + (94.5e^{-\tau_1 s} + 81e^{-2\tau_2 s})s^2$$
$$+ 170.1e^{-(\tau_1+\tau_2)s}s + 119.07e^{-2\tau_1 s}). \tag{4.16}$$

Hence the stability analysis requires studying separately the two factors containing the delay terms. Here the root at $s = 0$ corresponds to the "velocity consensus" nature of the dynamics. The remaining roots will then determine the stability of the consensus point.

Without duplicating a similar analysis, we state that a repeated zero root does not arise in this example. The remainder of the analysis follows from Chap. 3 in terms of obtaining the stability boundaries, identifying the stable region(s) on the delay plane, and superimposing the results for each of the factors of the characteristic function.

The factor $5s + 9e^{-\tau_2 s}$ is independent of τ_1 and carries only the delay term τ_2. This problem is analogous to that in (3.9) with $k = 9/5$ and thus the system is stable for $\tau_2 \in [0, 5\pi/18)$. The factor containing both delays needs to be analyzed following the guidelines on two-delay problems covered in Chap. 3. This analysis leads to stability

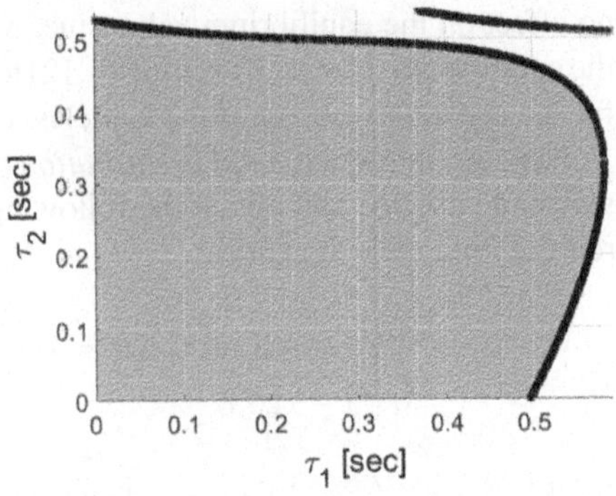

Fig. 4.4 Stability switching boundaries (not all shown) and stable region (gray shaded) for the dynamics of (4.13)–(4.14)

switching boundaries, one of which encapsulates the stable origin of the delay plane, as shown in Fig. 4.4, where stability is understood to be around the constant velocity rigid body dynamics. Here the region including the origin is consensus stable and the remaining boundaries are suppressed as they do not interfere with this region. Moreover, the stable region is well below the line $\tau_2 = 5\pi/18$; hence, the stability of the first factor in (4.16) does not affect the stability picture shown in Fig. 4.4.

4.4 Constant Time-Headway Driving Strategy

The mathematical modeling and the pertaining discussion in this section are adopted from Chaps. 1 and 2, as well as from [12]; however, the model is extended based on our interest in the effects of multiple delays. We start with the following model:

$$\dot{v}_k(t) = \alpha_k(v_{k+1}(t - \tau_{k+1}) - v_k(t - \tau_k)), \ k = 1, \ldots, n, \tag{4.17}$$

with all terms defined in the above presentation. Here we consider a different driving strategy, namely constant time-headway strategy, which assumes that driver k aims to perform driving actions to maintain a constant time-headway between vehicle k and $k + 1$ (see also [13] for details on different driving strategies). Specifically, the headway expression is given by

$$\delta_k(t) = x_{k+1}(t) - x_k(t) - l_k - \Delta_k - h_k v_k(t), \tag{4.18}$$

where $h_k > 0$ is the time-headway, and l_k and Δ_k are constants. Considering this driving strategy, we will investigate how the system stability is affected by delays τ_k and τ_{k+1}. To this end, we first demonstrate how this strategy can be tailored together with the model in (4.17).

4.4.1 Spacing Dynamics with Constant Time-Headway

Adopting from [12], let $\vartheta(s)$ denote the Laplace transform of $v(t)$. From (4.17) we obtain the transfer function between the velocities of consecutive vehicles as

$$G_k(s, \tau_k, \tau_{k+1}) := \frac{\vartheta_k(s)}{\vartheta_{k+1}(s)} = \frac{\alpha_k e^{-\tau_{k+1}s}}{s + \alpha_k e^{-\tau_k s}}. \tag{4.19}$$

Now differentiating the headway expression (4.18) and taking Laplace transforms with zero initial conditions, we can write in the Laplace domain

$$D_k(s) = \frac{1}{s}\left(\vartheta_{k+1}(s) - \vartheta_k(s)\right) - h_k \vartheta_k(s),$$

where $D_k(s)$ denotes the Laplace transform of $\delta_k(t)$. In view of (4.19), the above equation can be compactly expressed as

$$D_k(s) = \frac{1}{s}\left(1 - G_k - s h_k G_k\right) \vartheta_{k+1}(s).$$

Hence, the transfer function between two consecutive headways reads

$$\frac{D_k(s)}{D_{k+1}(s)} = \frac{1 - G_k - s h_k G_k}{1 - G_{k+1} - s h_{k+1} G_{k+1}} G_{k+1} := \hat{G}_k(s). \tag{4.20}$$

Since we have utilized (4.19), the effects of delays τ_k and τ_{k+1} are reflected in both velocity and position states of the time-headway driving strategy. In a similar manner, letting $V_k(s)$ and $A_k(s)$ denote the Laplace transforms of the velocity differences $v_{k+1}(t) - v_k(t)$ and the acceleration differences $\dot{v}_{k+1}(t) - \dot{v}_k(t)$, respectively, one can derive

$$\frac{V_k(s)}{V_{k+1}(s)} = \frac{A_k(s)}{A_{k+1}(s)} = \frac{1 - G_k}{1 - G_{k+1}} G_{k+1} = \bar{G}_k(s), \tag{4.21}$$

where some arguments are suppressed for easier reading. Notice that one recovers (4.21) from (4.20) by letting the $h_k \to 0$. Therefore, in the following, we only treat (4.20) for $h_k > 0$, and extensions to $h_k \to 0$ follow in a straightforward manner.

4.4.2 Preliminaries for Stability Analysis

From (4.20), we have

$$\frac{D_k(s)}{D_{k+1}(s)} = \frac{P(s, \overline{\tau}, \overline{\alpha})}{f(s, \overline{\tau}, \overline{\alpha})}, \tag{4.22}$$

where $\overline{\tau} = (\tau_k, \tau_{k+1}, \tau_{k+2})$ and $\overline{\alpha} = (\alpha_k, \alpha_{k+1})$, $f(s; \overline{\tau}, \overline{\alpha})$ is the characteristic function, given by

$$\begin{aligned} f(s; \overline{\tau}, \overline{\alpha}) =& s^2(-1 + h_{k+1}\,\alpha_{k+1}\, e^{-\tau_{k+2}\, s}) \\ &+ s\left(\alpha_{k+1}\, e^{-\tau_{k+2}\, s} - \alpha_{k+1}\, e^{-\tau_{k+1}\, s} - \alpha_k\, e^{-\tau_k\, s} + h_{k+1}\,\alpha_k \alpha_{k+1}\, e^{-(\tau_k+\tau_{k+2})s}\right) \\ &- \alpha_k \alpha_{k+1}\, e^{-\tau_k\, s}\left(e^{-\tau_{k+1}\, s} - e^{-\tau_{k+2} s}\right), \end{aligned} \tag{4.23}$$

and P in (4.22) can be extracted from (4.20).

The stability of the system can be analyzed starting with calculating the stability switching boundaries, for example, via the various approaches presented in the above examples. In the following, partly referring to Chap. 3, we present techniques to compute the stability switching boundaries. We then provide discussions on the results, which can be considered complementary to those obtained in this book on constant-headway driving strategies.

A critical point regarding (4.23) that is different from the characteristic functions treated so far is that this function is of *neutral* type, namely an exponential function with a delay term is now multiplying the highest order of s in $f(s, \overline{\tau}, \overline{\alpha})$. For this reason, additional considerations are needed [14–16]. Specifically, a necessary condition for the asymptotic stability of the system is that the delay-difference operator having the characteristic function

$$\tilde{f}(s, h_{k+1}, \alpha_{k+1}, \tau_{k+2}) = -1 + h_{k+1}\,\alpha_{k+1}\, e^{-\tau_{k+2}\, s} \tag{4.24}$$

is stable (here $\tilde{f}$ is the coefficient of the s^2 term in (4.23)). This condition in this particular problem is given by $|h_{k+1}\alpha_{k+1}| < 1$, since otherwise (4.24) has unstable zeros. If this condition is not satisfied, then the system associated with (4.23) is unstable for all positive delays τ_{k+2} even if the delay-free system is stable. Assuming that the above necessary condition is satisfied, we proceed to obtain the stability switching curves.

4.4.3 Stability Switching Boundaries

Recall that the detection of imaginary zeros, $s = \omega i$, of the characteristic function is the starting point in calculating the stability switching boundaries,

$$f(\omega i, \overline{\tau}, \overline{\alpha}) = 0, \tag{4.25}$$

where driver aggressiveness coefficients $\overline{\alpha}$ are assumed to be given. Notice that a "delay decoupling" property exists in (4.25):

$$f(\omega i, \overline{\tau}, \overline{\alpha}) = P(\omega i, \tau_k)\, Q(\omega i, \tau_{k+1}, \tau_{k+2}) = 0, \tag{4.26}$$

where

$$P = i\omega + \alpha_k e^{-i\omega\tau_k}, \tag{4.27}$$
$$Q = i\omega(h_{k+1}\alpha_{k+1}e^{-i\omega\tau_{k+2}} - 1) + \alpha_{k+1}(e^{-i\omega\tau_{k+2}} - e^{-i\omega\tau_{k+1}}). \tag{4.28}$$

Let us first study the stability of the delay-free system. When $\tau_k = \tau_{k+1} = \tau_{k+2} = 0$, the characteristic function has two roots $s_1 = 0$ and $s_2 = -\alpha_k < 0$. Owing to s_1 being an invariant root, the occurrence of a double zero root is also possible, similar to the case in Sect. 4.3.1. The system has a double zero pole for delay pairs $(\tau_{k+1}, \tau_{k+2}) \in \mathbb{R}_+^2$ that reside on the line defined by $L(\tau_{k+1}, \tau_{k+2}) = 1 + \alpha_{k+1}(\tau_{k+2} - \tau_{k+1} - h_{k+1}) = 0$. Moreover, as per the manipulation in (4.27)–(4.28) since the effects of τ_k are decoupled from those of (τ_{k+1}, τ_{k+2}), the following observations are immediate:

(a) Stability switching boundaries and the stability regions on the plane of (τ_{k+1}, τ_{k+2}) do not depend on τ_k.
(b) Imaginary solutions $s = \omega i$ of (4.26) along τ_k axis are give by $\tilde{\tau}_k = (\pi + \mp 4\pi\ell)/(2\alpha_k)$, $\ell = 0, 1, \ldots$.

Here, (a) is as per the decoupling effects, but the stability of the system can still be affected by τ_k since this term appears in one of the factors (4.27) of the characteristic function (4.26). Observation (b) is a consequence of the stability properties of (3.9), with the delay margin for factor (4.27) being $\tau_k^* = \pi/(2\alpha_k)$.

Consider next that $|h_{k+1}\alpha_{k+1}| < 1$ and $0 \leq \tau_k < \pi/(2\alpha_k)$. Then the dynamics around the rigid body motion is stable for any delay $\tau_{k+2} \in \mathbb{R}_+$ along the axis $\tau_{k+1} = 0$. A proof follows by setting $\tau_{k+1} = 0$, noticing that

$$|e^{-i\omega\tau_{k+2}}| < \left|\frac{\alpha_{k+1} + i\omega}{\alpha_{k+1}(i\omega h_{k+1} + 1)}\right|, \quad \forall\omega > 0,$$

and ignoring the rigid body motion at $\omega = 0$. Due to this inequality, it follows that Eq. (4.28) has no solutions τ_{k+2} for which a stability transition at $s = \omega i$ is possible. Since for zero delays the dynamics around the rigid body motion is stable, the dynamics remains stable along the τ_{k+2}-axis when $\tau_{k+1} = 0$. This result, in view of hyperbolicity (Sect. 3.3), indicates that we will observe a delay-independent stability feature along the τ_{k+2}-axis.

We now turn our attention to the stability switching boundaries for nonzero delays. Owing to the decoupling property, these boundaries can be studied from each factor $P(s, \tau_k) = 0$ and $Q(s, \tau_{k+1}, \tau_{k+2}) = 0$ separately.

Property 4.1 *Assume $|h_{k+1}\alpha_{k+1}| < 1$ holds, then the maximum of ω for which $s = i\omega$ is a solution to (4.25) is upper-bounded [14]. A conservative upper bound denoted by $\overline{\omega}$ is computed next. For calculating the exact upper bound, see, for example, [17].*

First, write

$$Q = Q_k(\omega i) + Q_{k+1}(\omega i)e^{-i\omega\tau_{k+1}} + Q_{k+2}(\omega i)e^{-i\omega\tau_{k+2}},$$

where $Q_k(\omega i) = -\omega i$. For $\omega \neq 0$, define $a_i = Q_{i+1}/Q_k$, $i = k, k+1$, and let Ω be the set of crossing frequencies. Then we have the following.

Theorem 4.1 *We have $\omega_0 \in \Omega$, if one of the following conditions is satisfied:*

(i) $\omega_0 = |\alpha_k|$;
(ii) The following triangle inequalities hold simultaneously:

$$|a_k(\omega_0 i)| + |a_{k+1}(\omega_0 i)| \geq 1 \quad \textit{and} \quad -1 \leq |a_k(\omega_0 i)| - |a_{k+1}(\omega_0 i)| \leq 1.$$

Remark 4.1 The behavior of the imaginary root crossing for $\omega = 0$ is characterized by the line $L(\tau_{k+1}, \tau_{k+2})$.

The analysis related to the classification of the stability crossing curves, smoothness properties, and the characterization of crossing directions follows closely the ideas presented in [18–20] and in Chap. 3.

Remark 4.2 In [12], the authors propose $\alpha_{k+1} = 0.37$ and $h_{k+1} = 1.8$ for human drivers. Although the cited work does not consider multiple delays, these numerical choices can be incorporated into the inequality conditions obtained within the above developed framework. If the traffic flow dynamics is governed by multiple delays, the condition $|h_{k+1}\alpha_{k+1}| = 0.666 < 1$ is not violated by these numerical choices. Moreover, one calculates the delay margin to be $\tau_k^* = \pi/(2\alpha_k) = 4.24$ s.

Remark 4.3 An alternative approach to compute the stability switching boundaries is to utilize the property that the exponential terms at $s = \omega i$ describe a unit circle on the complex plane. Let $s = \omega i$, and write the exponential functions in (4.23) as

$$e^{-\tau_m s} := x_m + iy_m, \quad m = k, k+1, k+2, \tag{4.29}$$

where x_m, y_m are real numbers. This substitution generates a new characteristic function $g(\omega i, x_m, y_m, \overline{\alpha}) = 0$ and conditions

$$\mathscr{C}_m := x_m^2 + y_m^2 - 1 = 0, \quad m = k, k+1, k+2. \tag{4.30}$$

Then the computation of stability switching boundaries reduces to simultaneously solving for the zeros of the circle equations and the characteristic function [21, 22].

Numerical Implementation. Extracting the stability switching boundaries:

- Given τ_k, if $|h_{k+1}\alpha_{k+1}| < 1$ and $0 \leq \tau_k < \pi/(2\alpha_k)$ hold, then go to the following step, else the system is unstable.
- Scan ω within a prescribed range as per Property 4.1.
- Follow the ideas in Chap. 3, or either one of the following:

Approach 1	Approach 2
Given h_{k+1} and $\overline{\alpha}$ values, solve $g(\omega i, x_m, y_m, \overline{\alpha}) = 0$ simultaneously with the unit circles $\mathscr{C}_{k+1}$ and $\mathscr{C}_{k+2}$. Common roots along with ω in (4.29) will yield the delay values τ_{k+1} and τ_{k+2}.	Use ω to compute the triangle inequalities in Theorem 4.1. If these inequalities are satisfied, one can use the arguments of the vectors $\|a_k\|$ and $\|a_{k+1}\|$ to compute the corresponding delays τ_{k+1} and τ_{k+2}.

- Repeat the above steps for $\omega \in (0, \overline{\omega})$ and calculate all the pairs (τ_{k+1}, τ_{k+2}). As per periodicity of the characteristic roots on the imaginary axis, compute additional delay pairs by shifting them with the phase angles $2\pi\ell/\omega$, $\ell = 0, 1, \ldots$. All these delays are on the stability switching curves (see Chap. 3).
- Given h_{k+1} and $\overline{\alpha}$, superimpose the line $L(\tau_{k+1}, \tau_{k+2}) = 0$ on top of the stability switching curves obtained in the previous step.
- Following Chap. 3 and the references therein, identify the regions where the system is either stable or unstable on the plane of delays τ_{k+1}–τ_{k+2}.

4.4.4 *Illustrative Examples*

Consider a chain of vehicles where each driver follows only the one in front. From this chain, we select a generic pair of drivers where the driver k is following the driver $k + 1$. In the first case, the two consecutive drivers have[4] $\alpha_k = 0.3$, $\alpha_{k+1} = 0.45$, and $h_k = h_{k+1} = 1.6$. The stability map of the vehicle-to-vehicle spacing dynamics is depicted in Fig. 4.5, where the region containing the origin corresponds to stable operation, keeping in mind that the stability region is valid only for $0 \leq \tau_k < 5.24$. When $\tau_k \geq 5.24$, the dynamics is unstable regardless of the values of τ_{k+1}, τ_{k+2}. Importantly, we remark that stability does *not* necessarily imply *collision-free* traffic flow, but indicates that perturbations against the spacing dynamics will be dampened out.

Next, we investigate stability for $\alpha_k = \alpha_{k+1} = 0.45$ and three different values of h_{k+1}, namely, 1.4, 1.6 and 1.8. We compare the corresponding stability regions in Fig. 4.6, left panel. Indeed, $h_{k+1} = 1.8$ is within the margins of practical time-headway values, whereas $h_{k+1} < 1.8$ could correspond to either a more aggressive (reactive) driver, or an automated vehicle. Therefore, in some sense, the stability regions in Fig. 4.6 compare the effectiveness of human drivers with more or less

[4] These numerical quantities are selected around the values $\alpha_k = 0.37$ and $h_k = 1.8$, proposed in [12]. Examples for different settings can be found in [23].

Fig. 4.5 Asymptotic stability regions for $\alpha_k = 0.3$, $\alpha_{k+1} = 0.45$, $h_{k+1} = 1.6$. For stability, $0 \leq \tau_k < 5.24$

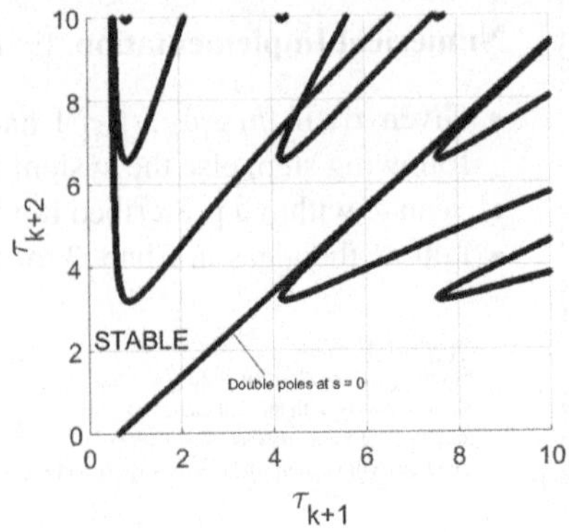

reactivity and of automated vehicles. We observe that stability region (the region containing the origin) is narrower and smaller with less reactive drivers (larger h_{k+1}), for example around the point $\tau_{k+1} = 1$ and $\tau_{k+2} = 4$.

Another scenario we look into is what happens if a more aggressive driver follows a less aggressive driver (Scenario 1), or vice versa (Scenario 2). For this case, we let $h_{k+1} = 1.6$ for both drivers in both scenarios. In Scenario 1, we set $\alpha_k = 0.45$ and $\alpha_{k+1} = 0.35$; and in scenario 2, we have $\alpha_k = 0.35$ and $\alpha_{k+1} = 0.45$. Figure 4.6-right panel shows the stability boundaries for Scenario 1 and Scenario 2 by thick and thin lines, respectively. Clearly, the thick curves yield larger space for the stability regions. That is, in the case when a more aggressive driver follows a less aggressive driver (Scenario 1), stability regions are relatively larger, indicating a larger degree of robustness against detrimental effects of delays.

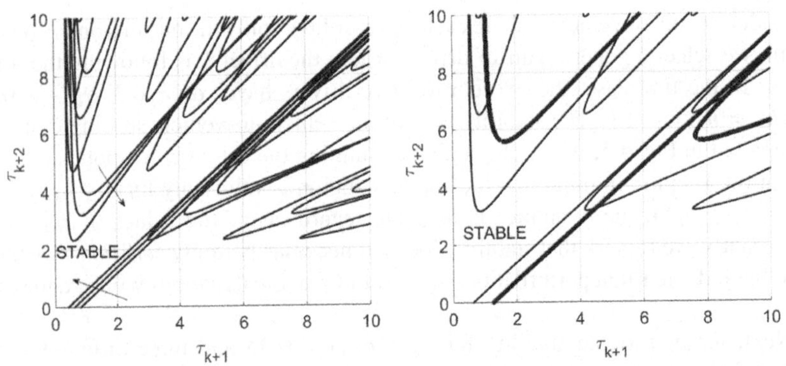

Fig. 4.6 Left: Constant time headway h_{k+1} affects stability where $\alpha_k = \alpha_{k+1} = 0.45$ and arrows show the order of $h_{k+1} = 1.4$, $h_{k+1} = 1.6$, $h_{k+1} = 1.8$. Here, for stability, $0 \leq \tau_k < 3.49$ must hold. Right: A more aggressive driver following a less aggressive driver (Scenario 1: thick black curves and $0 \leq \tau_k < 3.49$), or vice versa (Scenario 2: thin black curves and $0 \leq \tau_k < 4.49$) and how stability region is affected. Constant time-headway controller $h_{k+1} = 1.6$ for both drivers

4.5 Effects of Position Feedback

4.5.1 Mathematical Model

The above treatment focused on the effects of velocity feedback only. In this section, we investigate how the incorporation of position feedback could influence the interpretations. Specifically, we consider a position feedback only and a combined position-velocity feedback cases. Similar to the stability analysis performed in Sect. 4.2, here we consider $n = 3$ vehicles are arranged around a circular path and carry over the numerical quantities $\alpha_{k,1} = \gamma = 1.2$ and $\alpha_{k,2} = \beta = 0.75$. The driving strategies considered are as follows:

Drivers follow two vehicles ahead using *only position feedback*, i.e.,

$$\ddot{x}_k(t) = \sum_{j=1}^{2} \alpha_{k,j}(x_{k+j}(t-\tau_j) - x_k(t-\tau_j)). \tag{4.31}$$

Drivers follow two vehicles ahead using a *combination of position and velocity feedback*, i.e.,

$$\ddot{x}_k(t) = \sum_{j=1}^{2} \big(\alpha_{k,j}(x_{k+j}(t-\tau_j) - x_k(t-\tau_j)) + \chi_{k,j}(\dot{x}_{k+j}(t-\tau_j) - \dot{x}_k(t-\tau_j))\big). \tag{4.32}$$

Moreover, for the example study, we let the numerical value of $\chi_{k,j}$ in (4.32) to be equal to that of $\alpha_{k,j}$, noting, however, that these parameters have different units.

4.5.2 Stability Analysis Results

Position feedback only: In this case, it is easy to check that the delay-free dynamics is unstable for any selection of time delays, τ_1 and τ_2; see also [1]. We therefore do not further investigate this case.

Velocity and position feedback combined: For the dynamics in Eq. (4.32), the region connected to the origin of the delay plane is the stability region; see Fig. 4.7. This region, different from Fig. 4.2, is *bounded* on both axes. Consider next the assumption that $\tau_2 > \tau_1$. In this case, if a driver can have a delay less than $\tau_1 = 0.36$ s for sensing the velocity and position variations with respect to the vehicle immediately in the front, then the driver can have larger stability margins along τ_2 for sensing the velocity

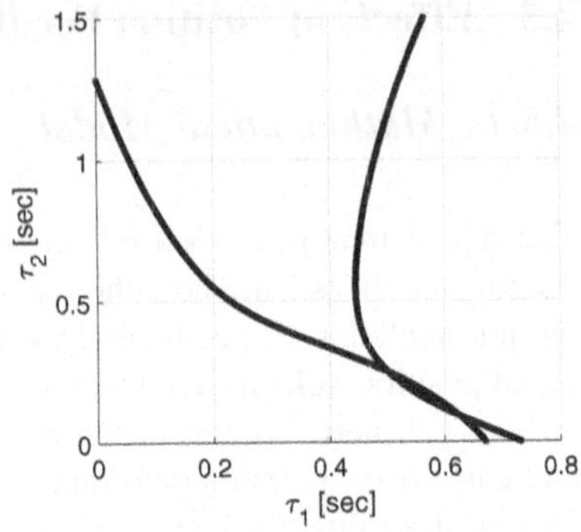

Fig. 4.7 Stability boundaries under velocity and position feedback (4.32). Stability region is connected to the origin of the delay plane

and position variations with respect to vehicle further ahead. Nonetheless, there is a trade-off between the delay margins, either one of them gets narrower while the other one enlarges, similar to the observations made in the previous example with velocity feedback only. Finally, we note that the delay-independent stability feature found in the case of velocity feedback only (see (4.6) and Fig. 4.2) is no longer available, but without position feedback it can be challenging to maintain a safe headway.

Overall, we find that velocity-only feedback control law is preferable since it renders the non-delayed dynamics stable. Moreover, one finds that the dynamics with $\tau_1 = 0$ is stable independent of τ_2. These discussions relate to weak/strong delay interactions as summarized in Chap. 3. On the other hand, an addition of position feedback is critical as this would be the only way vehicles can perform proper headway keeping.

4.6 Notes and Further Reading

It is important to keep in mind that the above analysis holds under certain assumptions. Firstly, the discussion assumes that the linear model used is representative of human driving. While there is evidence that such simple models can indeed be utilized to study human drivers (see, e.g., [12]), variability across drivers and for a driver under different conditions must be accounted for. Secondly, the considered model assumes time-invariance. Some previous studies discussed that human reaction delays may be assumed constant or slowly time-varying; however, here we assumed that the driver behavior captured by the aggressiveness coefficients β and γ are also constants. Last but not least, human behavior when following multiple vehicles may become anticipatory. For example in [24] human anticipation of upcoming traffic was modeled as part of the driving "strategy", and predictions by simulating the model showed that stability characteristics of traffic flow can indeed be enhanced in the case when humans take anticipatory driving decisions by following multiple vehicles ahead of them.

We have the following observations on the above obtained results:

1. As the case studies show, following multiple vehicles may not always be advantageous in the linear dynamics from a delay margin point of view, compared to following the preceding vehicle only. This is in part due to multiple-vehicle following causing high gains in the system, and one possibility to moderate this would be to scale the gains by the number of vehicles followed [25].
2. Stability may not be achievable only with position feedback control law. A larger stability margin may be possible for velocity feedback only, as opposed to velocity and position feedback together (compare Figs. 4.2 and 4.7). Furthermore, the region where the system is stable independently of τ_2 disappears with the additional position feedback terms in Eq. (4.32), and stability margin becomes finite, $0 \leq \tau_2 < 1.28$ s, when $\tau_1 = 0$.
3. Dynamics under certain control laws may exhibit multiple stability intervals as discussed in Chap. 3. This especially carries importance for controller implementation since it offers more options in the time-delay domain for the system to stably operate. In the presented traffic flow examples, such a feature was not observed.
4. Considering more realistic models involving multiple time delays may reveal interesting characteristics of the dynamics, and be a better representation of the physics of the problem. Moreover, with such considerations, unbounded stability regions may exist, or stability regions may enlarge in certain orientations in the parameter space.
5. Multiple delays can be regarded as multiple parameters affecting the dynamics, and the influence of one delay on stability may not be in the same sense for another delay parameter [26]. For example, an increase in one delay or a decrease in another can favor stability. From this point of view, there are a number of opportunities in studying the interactions of multiple vehicles under multiple sensing delays, and this know-how may be useful for future studies, such as in the design of interconnected vehicles.

With regard to Sect. 4.3.1, an interesting result in Fig. 4.3 is that for $0 < \tau_1 < 0.27$ s the dynamics becomes stable independently of τ_2. This result concludes that, regardless of the delay τ_2 of a driver's position sensing of the preceding vehicle, the linear traffic dynamics in Eq. (4.8) is guaranteed to be stable as long as the drivers act on their own positions with delays no more than 0.27 s.

Furthermore, a comparison of taking different ratios for τ_2/τ_1 in Fig. 4.3 reveals enhancement of stability margins along these ratios. For example, consider first the conventional preference of $\tau_2/\tau_1 = 1$, which recovers the single delay problem. Then, the maximum allowable time-delay maintaining stability would be $\tau_{max,1} \approx 0.41$ s. On the other hand, if one assumes that sensing delays of drivers about the position of their preceding vehicle is three times longer than that of their own position, i.e., $\tau_2/\tau_1 = 3$, the maximum allowable time-delay maintaining stability would be $\tau_{max,1} = 0.3$ s and $\tau_{max,2} = 0.9$ s. That is, with a 25% drop in τ_1, it becomes possible to expand system's delay tolerance along the axis of τ_2 by more than a factor of two.

There also exist various scenarios in the way that the dynamics becomes oscillatory and unstable, as characterized by Fig. 4.3: (a) If $\tau_2 = 0$ and τ_1 is increased beyond $\tau_1 = 0.54$ (crossing the double root boundary), a dominant behavior of exponential increase in system states is observed, due to the crossing of a single root $s = 0$ across the origin toward the positive real axis of the complex plane. (b) Selection of τ_1 and τ_2 on the boundaries of the stability region where the characteristic function has a pair of imaginary zeros causes an oscillatory behavior of the states with fixed amplitudes in steady state. (c) Selection of τ_1 and τ_2 outside of the stability region in general causes a dominant behavior of oscillations of states with increasing amplitudes. (d) Selection of τ_1 and τ_2 outside the stability region that is nearby the double root boundary and the rest of the boundaries would cause a combination of exponentially increasing behavior and oscillatory behavior with increasing amplitudes, i.e., a linear combination of the effects observed in (a) and (c) above.

With regard to Sect. 4.3.2, we note that the stability region in Fig. 4.4 has an interesting geometry, especially when the ratio $\tau_2/\tau_1 \in (1/2, 1)$, where the stability region bulges and enhances. This improves the stability region, which we attribute to the *strong interaction between the delays*. For further insight, consider a delay ratio of $\tau_2/\tau_1 = 3/4$ in Fig. 4.4, which yields a delay margin of $\tau_{\max,1} \approx 0.5563$ s and $\tau_{\max,2} \approx 0.4176$ sec, compared to the margin of 0.47 s obtained for the ratio of unity. Interestingly, these margins are comparable and in some cases larger than the margin of 0.496 obtained by assuming that $\tau_2 = 0$. This is potentially another demonstration of how delay uncertainties may work in favor of stability, or vice versa, as they may shift the delay ratios and hence affect the delay margin. Ultimately, looking into the problem as a two-delay problem instead of a single-delay one, based on the motivation that there may be a delay mismatch and delay uncertainty in the system, provides a wider perspective regarding how different delays interact with each other and affect overall system stability.

References

1. R. Sipahi, S.-I. Niculescu, Analytical stability study of a deterministic car following model under multiple delay interactions. IFAC Proceedings Volumes, **39**(10), 187–192 (2006)
2. R. Sipahi, S.-I. Niculescu, Some remarks on the characterization of delay interactions in deterministic car following models, 17th International Symposium on Mathematical Theory of Networks and Systems (MTNS), 2006, Japan
3. W. Qiao, R. Sipahi, Delay-dependent coupling for a multi-agent LTI consensus system with inter-agent delays. Phys. D: Nonlinear Phenom. **267**, 112–122 (2014)
4. W. Qiao, R. Sipahi, A linear time-invariant consensus dynamics with homogeneous delays: analytical study and synthesis of rightmost eigenvalues. SIAM J. Control Optim. **51**(5), 3971–3992 (2013)
5. W. Qiao, R. Sipahi, Rules and limitations of building delay-tolerant topologies for coupled systems. Phys. Rev. E **85**(2), 016104 (2012)
6. W. Qiao, R. Sipahi, Responsible eigenvalue concept for the stability of a class of single-delay consensus dynamics with fixed topology. IET Control Theory Appl. **5**(4), 622–629 (2011)
7. R. Sipahi, F.M. Atay, S.-I. Niculescu, Stability of traffic flow with distributed delays modeling the memory effects of the drivers. SIAM Appl. Math. **68**(3), 738–759 (2008)

8. R. Olfati-Saber, R.M. Murray, Consensus problems in networks of agents with switching topology and time-delays. IEEE Trans. Autom. Control **49**, 1520–1533 (2004)
9. R.E. Chandler, R. Herman, E.W. Montroll, Traffic dynamics: studies in car following. Oper. Res. Int. J. **6**(2), 165–184 (1958)
10. M. Bando, K. Hasebe, K. Nakanishi, A. Nakayama, Analysis of optimal velocity model with explicit delay. Phys. Rev. E **58**, 5429–5435 (1998)
11. R. Sipahi, *Cluster Treatment of Characteristic Roots, CTCR, A Unique Methodology for the Complete Stability Robustness Analysis of Linear Time Invariant Multiple Time Delay Systems Against Delay Uncertainties*, Ph.D. Dissertation, University of Connecticut, Mechanical Engineering Department (2005)
12. A. Bose, P.A. Ioannou, Analysis of traffic flow with mixed manual and semi automated vehicles. IEEE Trans. Intell. Transp. Syst. **4**(4), 173–188 (2003)
13. M. Treiber, A. Kesting, *Traffic Flow Dynamics* (Data, Models and Simulation, Springer, Berlin, Heidelberg, 2013)
14. J. K. Hale, S.M. Verduyn Lunel, *Introduction to Functional Differential Equations* (Springer, New York, 1993)
15. H. Logemann, S. Townley, The effect of small delays in the feedback loop on the stability of neutral systems. Syst. Control Lett. **27**, 267–274 (1996)
16. W. Michiels, K. Engelborghs, D. Roose, D. Dochain, Sensitivity to infinitesimal delays in neutral equations. SIAM J. Control Optim. **40**, 1134–1158 (2002)
17. I.I. Delice, R. Sipahi, Delay-independent stability test for systems with multiple time delays. IEEE Trans. Autom. Control **57**(4), 963–972 (2011)
18. K. Gu, S.-I. Niculescu, J. Chen, On stability crossing curves for general systems with two delays. J. Math. Anal. Appl. **311**, 231–253 (2005)
19. R. Sipahi, N. Olgac, Complete stability robustness of third-order LTI multiple time-delay systems. Automatica **41**, 1413–1422 (2005)
20. R. Sipahi, *Mastering Frequency Domain Techniques for the Stability Analysis of LTI Time Delay Systems*, vol. 20 (SIAM, Philadelphia, 2019)
21. R. Sipahi, N. Olgac, *Stability Map of Systems with Three Independent Delays* (American Control Conference, Minneapolis, Minnesota, 2006)
22. R. Sipahi, I.I. Delice, Extraction of 3D stability switching hypersurfaces of a time delay system with multiple fixed delays. Automatica **45**, 1449–1454 (2009)
23. R. Sipahi, S.-I. Niculescu, I.I. Delice, Asymptotic stability of constant time headway driving strategy with multiple driver reaction delays, *American Control Conference, St Louis* (2009)
24. M. Treiber, A. Kesting, D. Helbing, Delays, inaccuracies and anticipation in microscopic traffic models. Phys. A **360**(1), 71–88 (2006)
25. D. Wang, R. Sipahi, Stability of a large-scale connected vehicle network in ring configuration and with multiple delays. IEEE Trans. Intell. Transp. Syst. **23**(1), 663–667 (2020)
26. N. MacDonald, Two delays may not destabilize although either delay can. Math. Biosci. **82**, 127–140 (1986)

Chapter 5
Linear Stability of Traffic Flow Models with Distributed Delays

This chapter explores the modeling of drivers' memory using distributed delays and demonstrates how such delays can be incorporated into car-following models. To this end, a benchmark model is taken to present frequency domain ideas, with the aim of analyzing the stability of the dynamics with respect to driver memory parameters and drivers' sensitivity to speed corrections. The analysis takes advantage of an interesting decoupling property of the system characteristic equation, allowing a simple and transparent interpretation.

From a practical point of view, the results obtained from the stability analysis are utilized to interpret how drivers' aggressiveness, as measured by their sensitivity to speed corrections, would affect stability and particularly the tolerable limits describing the distributed delays. This is especially important when taking into account the heterogeneity of the drivers and how it influences the traffic flow dynamics in the presence of delays. Furthermore, the effect of network structure on the stability is discussed. A specific interest is to understand the stability features when the vehicles are arranged in a chain versus ring configuration. The chapter closes with a consideration of a mix of drivers with more and less aggressive driving characteristics, and how this mixing impacts flow stability.

5.1 Driver Memory Effects Modeled by Distributed Delays

The starting point in this chapter is the model given in (2.2):

$$\dot{v}_k(t) = \alpha_k \int_0^\infty f_k(\tau)(v_{k+1}(t-\tau) - v_k(t-\tau))\, d\tau, \quad k = 1, \ldots, n, \tag{5.1}$$

where v_k is the velocity of the kth vehicle, and $\alpha_k > 0$ can be considered as the aggressiveness or the sensitivity of the kth driver in correcting the velocity

R. Sipahi et al., *Deterministic Car-Following Traffic Models*,
SpringerBriefs in Control, Automation and Robotics,
https://doi.org/10.1007/978-3-031-58164-9_5

discrepancies between the two vehicles. See also [1] and the explanations following (2.2) for further details.

Equation (5.1) corresponds to a linear vehicle model whose behavior is coupled with that of the preceding vehicle and whose acceleration decisions are governed by a continuous collection of information from past, mimicking a memory effect. Furthermore, a more realistic memory model should limit the capacity of the memory, i.e., the size of the memory window must be finite, which can be adjusted by a proper selection of the delay kernel $f_k(\tau)$.

For simplicity of presentation, assume that the delay kernel is homogeneous for all the drivers, i.e., $f_k(\tau) = f(\tau)$ for all k. Then, the system in (5.1) is expressed as

$$\dot{v}(t) = \int_0^{\infty} Jv(t-\tau)f(\tau)\,d\tau, \tag{5.2}$$

where $v = (v_1, \ldots, v_n)^T$ and $J \in \mathbb{R}^{n\times n}$ is the configuration matrix weighted by driver sensitivities α_k. Next, two spatial configurations are considered for the vehicles: (a) arranged along a linear path (Fig. 1.2), and (b) around a ring (Fig. 1.1). In the ring configuration, we have $v_0 = v_n$ in (5.1), hence J reads

$$J = \begin{pmatrix} -\alpha_1 & 0 & \cdots & 0 & \alpha_1 \\ \alpha_2 & -\alpha_2 & 0 & \cdots & 0 \\ 0 & \alpha_3 & -\alpha_3 & \cdots & 0 \\ \vdots & \vdots & \ddots & \ddots & \vdots \\ 0 & \cdots & 0 & \alpha_n & -\alpha_n \end{pmatrix}. \tag{5.3}$$

In the linear configuration, the vehicle at the very front of the platoon, denoted with the index $k = 1$, is the leader, traveling at a constant velocity, i.e., $\dot{v}_1 = 0$, which is independent of the dynamics of the remaining vehicles. In this case, J can be recovered from the ring configuration simply by setting $\alpha_1 = 0$.

Next, we analyze the stability of (5.2) for a given delay kernel $f(\tau)$, and investigate how linear versus ring configurations as well as drivers' aggressiveness represented by α_k affect stability. To this end, we first develop the necessary stability analysis tools, primarily following from Chap. 3.

5.2 Distributed Delays and Characteristic Functions

5.2.1 *Delay Distributions*

We examine the effects of two distribution functions $f(\tau)$: uniform distribution and γ-distribution with/without gap.

5.2.1.1 Uniform Distribution

The uniform distribution is given by the density function

$$f(\tau) = \begin{cases} 1/\delta \text{ if } h < \tau < h + \delta \\ 0 \quad \text{otherwise,} \end{cases} \tag{5.4}$$

where $\delta > 0$ and $h \geq 0$ are parameters. This kernel is a simpler first-order approximation of the short-term memory, corresponding to the average of the information available. It is of particular interest to understand how "old" the information can be allowed to be, to still be able to maintain stability. To this end, one investigation will be on the effects of dead-time h on stability, for a given memory size δ.

5.2.1.2 γ-distribution with and Without Gap

The gamma distribution, or γ-distribution, is given by the density function

$$f(\tau) = \begin{cases} \dfrac{(\tau - h)^{p-1} e^{-(\tau-h)/q}}{q^p \Gamma(p)} & \text{if } \tau \geq h, \\ 0 & \text{if } \tau < h, \end{cases} \tag{5.5}$$

where p, q are positive respectively describing the shape and scale parameters, Γ is the Gamma function, and the gap $h \geq 0$ is the dead-time. We have that f has a mean of $h + pq$, and its variance is given by pq^2. Here, the variance is a measure of the length of the memory window, where q is a scaling factor. It is of particular interest to find out how the parameters h, p, and q affect the stability of (5.2).

5.2.2 *Features of the Characteristic Function*

The characteristic function is obtained from (5.2) as

$$\chi(s, J) = \det[sI - JF(s)], \tag{5.6}$$

where $\chi : \mathbb{C} \times \mathbb{R}^{n \times n} \to \mathbb{C}$, and $F(s)$ is the Laplace transform of the distribution function $f(\tau)$. In the case of uniform distribution

$$F(s) = \frac{e^{-sh}(1 - e^{-s\delta})}{s\delta}. \tag{5.7}$$

In the limit $\delta \to 0^+$, one recovers the discrete delay model (2.1) with $\tau = h$. In the case of a γ-distribution with gap,

$$F(s) = \frac{e^{-hs}}{(qs+1)^p}, \tag{5.8}$$

which in the limit $h \to 0^+$ recovers a standard finite-dimensional transfer function. Observe that in general Eq. (5.6) can be rewritten as a multiplication of factors,

$$\chi(s, J) = \prod_{r=1}^{n} \chi_r(s, \lambda_r(J)),$$

where each factor $\chi_r : \mathbb{C} \times \mathbb{C} \to \mathbb{C}$ is given by

$$\chi_r(s, \lambda_r(J)) = s - \lambda_r(J)F(s), \tag{5.9}$$

with $\lambda_r(J)$ being the rth eigenvalue of J. The roots of χ_r are the poles of the system, which are parameterized with respect to the eigenvalues of J and determine the stability of the system.

As we have seen in the examples in Chap. 4, the configuration matrix J has a zero eigenvalue $\lambda_1 = 0$, corresponding to the rigid body motion of the ensemble of vehicles; furthermore, under mild connectivity conditions, zero is a simple eigenvalue and the remaining eigenvalues λ_r, $r = 2, \ldots, n$, have negative real parts [1]. Since $\lambda_1 = 0$, the first factor $\chi_1(s) = s$ has a root at $s = 0$, which can be shown to be a simple root of χ [2]. Since the simple zero root at $s = 0$ has no influence on the stability of the dynamics, in the remainder of this chapter we will focus only on the nonzero eigenvalues of J.

In the case of γ-distribution with gap, the arising characteristic function $\chi_r(s)$ takes a similar form as that in Chap. 3. Hence, the details of this analysis will not be repeated in this chapter and instead some illustrative examples will be provided. On the other hand, in the case of uniform distribution (5.7), there are a number of unique features in (5.9) that need further attention and require one to take a different approach to compute the stability switching boundaries of the dynamics. These technical points are highlighted as follows. First of all, note that (5.9) with (5.7) has two independent time-delay parameters, (h, δ). To this end, we seek for stability "regions" on the plane of (h, δ). Next, the delay δ not only appears in an exponential function, but also as a coefficient in the characteristic equation. This is a feature that was not explicitly covered in Chap. 3, but we can still provide a treatment following the main ideas in Chap. 3.

Remark 5.1 A traffic scenario can contain about one hundred vehicles (see Chaps. 1–2). However, for a large number of vehicles the analysis of the characteristic equation (5.6) becomes complicated. To address this issue, we can instead study the spectra of (5.9) with respect to each eigenvalue $\lambda_r(J)$, one by one, obtaining the stability of the system corresponding to each factor χ_r. We can then combine all these results next to find the ultimate stability region. Clearly, studying (5.9) with (5.7) is much simpler, since χ_r represents a lower dimensional dynamics. That is, the dimension of the problem is reduced from analyzing a high-order dynamics only once to *repeatedly* analyzing a low-dimensional dynamics for $n-1$ times, $r = 2, \ldots, n$. Such a reduction in order is quite beneficial when studying large-scale systems in general. Nevertheless, χ_r carries complex coefficients since $\lambda_r(J) \in \mathbb{C}_-$, making standard approaches not directly applicable in this case.

A remedy to the issues described above is the following. From (5.9), finding the nonzero roots of $\chi_r = 0$ is equivalent to finding the nonzero solutions s of the equation

$$\frac{\lambda_r(J)F(s)}{s} = 1.$$

The left-hand side can be factored in a judicious way to facilitate the analysis, say,

$$H_r(s) \cdot \Delta(s) = 1. \tag{5.10}$$

For the case of uniformly distributed delays, we have

$$H_r(s) = \frac{\lambda_r(J)e^{-sh}}{s} \quad \text{and} \quad \Delta(s) = \frac{1 - e^{-s\delta}}{s\delta}. \tag{5.11}$$

The idea of rewriting the problem at hand as in (5.10) is to enable the separation of the unknown parameters into either factor. The form in (5.10) is an interconnection scheme, which represents the characteristic function (5.9) in two separate blocks where each of the unknowns δ and h appear only in one block but not in the other. This decoupling ultimately enables an effective means to compute the stability switching curves.

Remark 5.2 It is easy to show that the interconnection scheme for the case of γ-distributed delay has the same $H_r(s)$ as in (5.11), but $\Delta(s)$ is different, given by

$$\Delta(s) = \frac{1}{(qs+1)^p}. \tag{5.12}$$

We continue the presentation on uniformly distributed delays.

5.2.3 An Algorithm for Obtaining Stability Switching Curves

Based on the continuity of system poles with respect to system parameters [1], it suffices to study system stability based on the stability switching curves, as explained in Chap. 4. For the specific problem at hand, check first whether or not the delay-free system (i.e., set $h = \delta = 0$ in (5.9)) is stable; determine all the imaginary roots $s = i\omega$ of the interconnection scheme in the domain of delays (h, δ); check the sensitivity of $s = i\omega$ with respect to parameters h and δ, in order to determine whether the roots move to the left- or to the right-half complex plane; and finally label as stable the regions in (h, δ) where the system has no unstable poles.

Specifically, set $s = i\omega$ in χ, then

$$H_r(i\omega) \cdot \Delta(i\omega) = 1. \tag{5.13}$$

One can then follow a geometric perspective as follows. The interconnection (5.13) can be studied separately for its magnitude and arguments on the complex plane. Without loss of generality, let $\omega > 0$, and notice that $|e^{-ih\omega}| = 1, \forall (h, \omega) \in \mathbb{R}^2$; thus an algorithm to compute h, δ pairs can be constructed as follows [1]:

Step 1: Given $\lambda_r(J)$. Let $z = \delta\omega/2$. To obtain the correspondence $\delta \leftrightarrow \omega$, one can consider z is a sweeping parameter, i.e., it is scanned within a range.
Step 2: Based on the magnitude condition, we have

$$\delta = \frac{2z^2}{|\lambda_r(J)||\sin(z)|}, \tag{5.14}$$

that is, δ can be computed as the parameter z is scanned.
Step 3: From the definition of z in Step 1 and since δ is known from Step 2, ω reads

$$\omega = \frac{|\lambda_r(J)||\sin(z)|}{z}. \tag{5.15}$$

Step 4: The argument condition on (5.13) yields

$$h = -\frac{1}{\omega}\left(\angle\left(\frac{\delta s^2}{\lambda_r(J)(1 - e^{-s\delta})}\Bigg|_{s=i\omega}\right) \pm 2\pi\ell\right), \quad \ell \in \mathbb{N}. \tag{5.16}$$

The above algorithm yields the triplets (ω, h, δ) and, next, we should reveal where on the (h, δ) plane the system is stable. The ideas in Chap. 3 can be leveraged to complete this task.

5.3 Illustrative Examples and Interpretations

We present several case studies for the stability analysis of the traffic dynamics (5.2), with distributed delays from a uniform distribution and a γ-distribution with gap. With uniform delay distribution, we consider the following scenarios: how vehicle configuration, in particular linear versus chain configurations, affects stability; how memory size is influenced by the drivers' aggressiveness parameter; how the existence of more aggressive drivers can affect stability. The scenarios with γ-distribution with gap include how the mean and variance of the distribution as well as the number of vehicles influence stability when the gap is set to zero and how nonzero gap changes these outcomes.

5.3.1 *The Case with Uniform Distribution*

We first study the effects of driver aggressiveness in linear and ring configurations. For this, let $n = 10$ and $\alpha_k = \alpha = 2.0$, in both configurations. This yields the stability characterization in Fig. 5.1, where stability region is the region below the computed stability switching boundaries. We find that the linear configuration of the vehicles yields a larger stability region $S_1 \cup S_2$. In Fig. 5.1, one can also identify $\max \delta = \bar{\delta}$ as the maximum achievable memory size. For example, in the case of linear configuration $\bar{\delta} \approx 2.5$. Moreover, there exists some degree of robustness against n and α as the boundaries are similar to those obtained for $n = 20$ in [1] (see below for further discussions).

It is shown in [1] that for the traffic flow model considered, the maximum achievable memory window size $\max(\delta) = \bar{\delta}$ occurs when the dead-time h is zero, for both the ring and linear configurations of the vehicles. Moreover, the computations

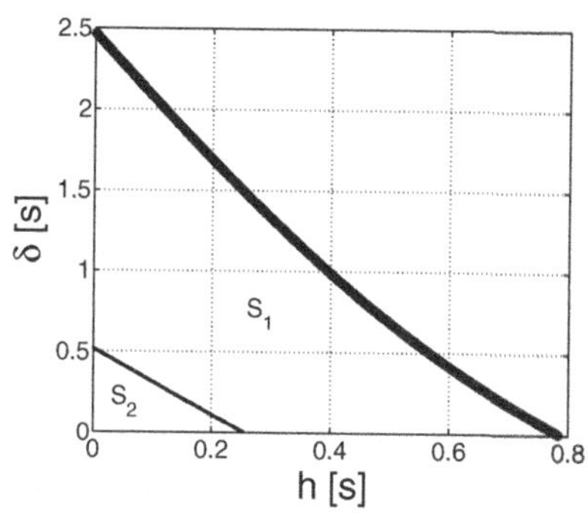

Fig. 5.1 Comparison of stability regions for linear and ring configurations with $n = 10$ vehicles and driver sensitivity $\alpha_k = \alpha = 2.0$. The thick and thin curves correspond to, respectively, the linear and ring configurations of vehicles. The union of the regions $S_1 \cup S_2$ is the stability region for the linear configuration, while the region S_2 is the stability region for the ring configuration

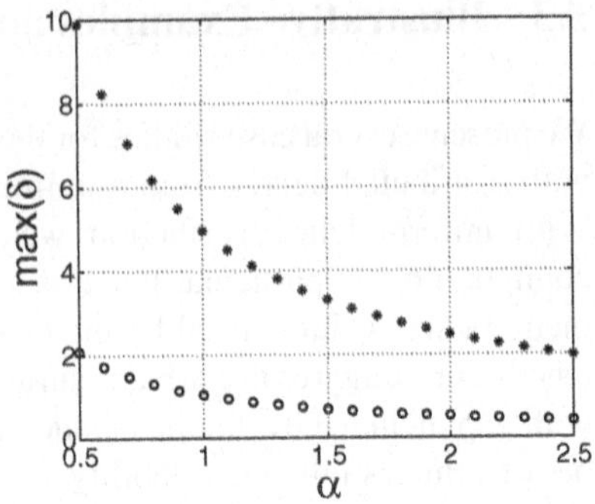

Fig. 5.2 Maximum achievable memory window size, $\max \delta = \bar{\delta}$ decreases with respect to increasing driver aggressiveness $\alpha_k = \alpha$. Here, we have $n = 10$ vehicles in the linear ($*$) and ring ($\circ$) configurations

suggest that the stability boundary, see Fig. 5.1, is monotonically decreasing as h is increased. To a certain degree, $\bar{\delta}$ can be used as a measure of tolerance to delays, and can thus be studied for further investigation. For example, in the presence of aggressive drivers (i.e., larger α), we find that the linear configuration of vehicles for the selected range of α allows a larger value of $\bar{\delta}$ for the drivers compared to the ring configuration. On the other hand, $\bar{\delta}$ is much more sensitive with respect to α in the linear configuration, Fig. 5.2.

We next investigate how the presence of aggressive drivers among less aggressive drivers could affect the stability analysis. To this end, some of the drivers, whose positions in the configuration are picked randomly, are assumed to be more aggressive, i.e., their α_k is increased by the same amount based on a certain percentage in the range of 5–30%. The remaining drivers are assumed to have identical α_k and the results are obtained in terms of how the maximum achievable memory window size $\bar{\delta}$ is affected with respect to added aggressiveness in percentage; see Fig. 5.3 for ring and linear configurations.

The results in Fig. 5.3 can be interpreted as follows. In the ring configuration, for a given percentage increase in aggressiveness, a larger number of aggressive drivers causes a larger drop in $\bar{\delta}$. Moreover, $\bar{\delta}$ consistently decreases with added percentages in driver aggressiveness. Intuitively, the results make sense as the presence of more aggressive drivers leads to smaller amounts of tolerable delays. Another observation is regarding the relative drop in $\bar{\delta}$, which is about 20% in both panels in Fig. 5.3. Furthermore, in the linear configuration, the decrease in the memory window size is independent of the number of aggressive vehicles among the ten vehicles. This is due to the eigenvalues λ_r of J in this configuration, which are such that the number of aggressive vehicles does not necessarily influence the particular eigenvalue that determines the stability regions.

We now turn our attention to the effect of number of vehicles on stability regions. Two different values for the number of vehicles are considered, with $n = 3$ and $n = 100$, both on the ring configuration and with two different driver aggressiveness values $\alpha = 1.0$ and $\alpha = 2.0$. Although $n = 3$ represents a small-size traffic, it is useful for comparison purposes. Following the framework explained above, one obtains the stability pictures in Fig. 5.4 depicted on the parameter plane of mean memory versus memory window size. In Fig. 5.4, the stability region when $n = 3$

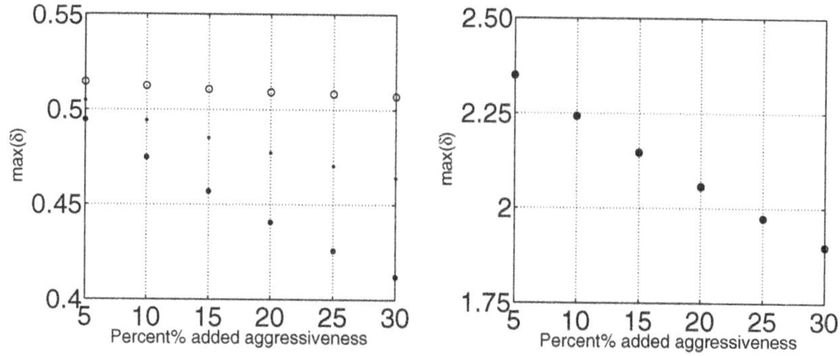

Fig. 5.3 Change in maximum achievable memory window size max $\delta = \bar{\delta}$ with respect to percentage of added aggressiveness to one (large circle), five (full dots), or nine (small circle) drivers, picked randomly, in a ten-vehicle scenario in the ring (left) and linear configurations (right). Data points for a given percentage are identical (right) hence overlap due to the specific vehicle configuration considered

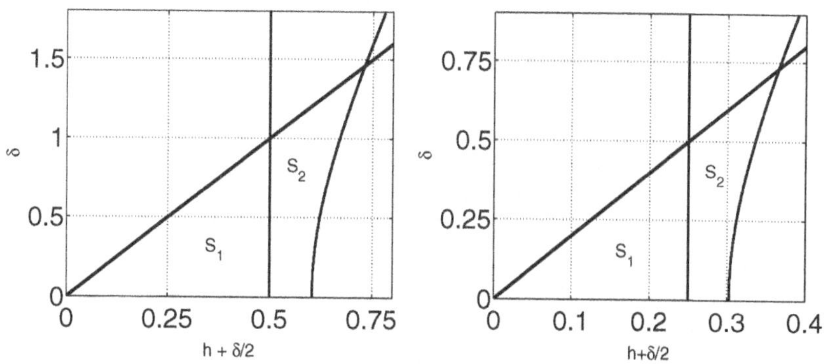

Fig. 5.4 Comparison of stability regions with respect to the mean of memory window $h + \delta/2$ versus memory window size δ. Driver aggressiveness is taken as $\alpha = 1.0$ (left figure) and $\alpha = 2.0$ (right figure) with $n = 3$ (stable region: $S_1 \cup S_2$) and $n = 100$ vehicles (stable region: S_1). See [3] for a stability map for $\alpha = 1.5$ and $n = 5$ and $n = 25$

is the union of the two regions S_1 and S_2 while the stability region shrinks to S_1 for $n = 100$. Although not depicted, we note that the reduction in the stability zone when n is further increased is negligible for $n > 100$. In some sense, for large n the stability region converges to a region very close to S_1. It is worthwhile to point out that the right-hand boundary of the stability regions is sensitive to small n, but as n grows this boundary shifts to the left only at a negligible amount, indicating a certain degree of stability robustness to n.

According to Fig. 5.4, when $n = 3$ and $\alpha = 1.0$, stability can be maintained even if the drivers utilize older information, such as for the selection of mean of 0.67 and memory window size of 1.1. This selection would however destabilize the dynamics for $n = 100$. In other words, as the scale of the problem is increased, the drivers should use more recent information in their memory for maintaining stability, instead of relatively older information.

5.3.2 Automatic Headway Controllers with Driver Memory

Automated inter-vehicle spacing control strategies have been widely studied and investigated, including sliding mode controllers [4, 5], proportional–derivative (PD) type controllers [6], adaptive fuzzy controllers [7], and lead controllers [8]. Below, we utilize PD controllers to demonstrate some ideas along these lines, but in the presence of driver memory [9].

Consider the following linear car-following model with memory effects and headway controller $u_k(t)$

$$\ddot{y}_k(t) = \alpha_k \int_0^\infty f_k(\tau)H_k(t-\tau)d\tau - b_k\dot{y}_k(t) + u_k(t), \tag{5.17}$$

$$H_k(t) = y_{k+1}(t) - y_k(t), \tag{5.18}$$

$$u_k(t) = K_{p,k}H_k(t) + K_{d,k}\dot{H}_k(t), \tag{5.19}$$

where H_k is the headway perturbations between the vehicles k and $(k+1)$ obtained by position perturbations $y_k(t)$ of the vehicles. In the model, perturbations due to drag forces are represented by the term $b_k\dot{y}_k(t)$, where b_k is the drag coefficient, and the gains $K_{p,k}$ and $K_{d,k}$ correspond to proportional and derivative controller gains per unit mass of the vehicle, respectively.

For demonstration purposes, in the sequel we choose the linear configuration for the vehicles, and assume that the memory effect is homogeneous across drivers and modeled by the uniform distribution (5.4). Notice in this case that, due to the linear arrangement of the vehicles, it is possible to consider the dynamics between two consecutive vehicles as representative of any of the two vehicles following each other. In the ring configuration, however, such a simplification is not possible in general, and hence one would need to express the state space representation of the entire dynamics for n vehicles.

With the simplification obtained based on the linear configuration, one can easily show that the Laplace transforms $Y_k(s)$ and $Y_{k+1}(s)$ of the perturbations $y_k(t)$ and $y_{k+1}(t)$, respectively, are related by the following transfer function:

$$G_k(s) := \frac{Y_k(s)}{Y_{k+1}(s)} = \frac{K_{p,k} + K_{d,k}s + \alpha_k F(s)}{s^2 + (K_{d,k}+b)s + K_{p,k} + \alpha_k F(s)}, \tag{5.20}$$

where $F(s)$ is given in (5.7) for the case of uniform distribution. We obtain the characteristic function of the dynamics in (5.20) as

$$f(s, h, \delta) = s^2 + (K_d + b)s + K_p + \alpha F(s), \tag{5.21}$$

where the subscript k is dropped as the above function applies to any two consecutive vehicles. In the following, we wish to reveal the parametric settings of the controller gains K_p and K_d for which system stability holds. To this end, we will pick the parameters of the memory distribution as well as the damping coefficient b.

For the analysis, it is important to determine the values of the controller gains for which the dynamics might be in transition from stability to instability (or, vice versa). In line with the discussions in Chap. 3, we study the zeros of the characteristic function at $s = i\omega$, that is,

$$f(s,h,\delta) = -\omega^2 + i\omega(K_d + b) + K_p + \alpha e^{i\omega h}\frac{1-e^{-i\omega\delta}}{i\omega\delta} = 0. \tag{5.22}$$

Separating the real and imaginary parts, we obtain

$$K_p = \omega^2 - \frac{2\alpha}{\omega\delta}\cos(\omega\tau_a)\sin\left(\frac{\omega\delta}{2}\right), \quad K_d = \frac{2\alpha}{\omega^2\delta}\sin(\omega\tau_a)\sin\left(\frac{\omega\delta}{2}\right) - b, \tag{5.23}$$

where $\tau_a = h + \delta/2$ is the mean of the memory distribution.

With the explicit formulae of the controller gains K_p and K_d available in terms of the sweep parameter ω, it is possible to compute these gains, which correspond to the characteristic function having roots on the imaginary axis. Notice that f has a zero when $s = 0$ and $K_p = -\alpha$. These gains will therefore form the potential points across which system stability may change.[1] The controller gains for $\delta = 1$, $h = 0.2$, and $b = 0.2$ with either $\alpha = 1$ or $\alpha = 2$ are obtained as shown in Fig. 5.5. In this figure, for a given α, two main curves $\mathscr{C}_1$ and $\mathscr{C}_2$ exist. The curve $\mathscr{C}_1$ is computed by (5.23), and $\mathscr{C}_2$ by observing that as $K_p \to -\alpha$ the system has a zero pole, creating a vertical boundary parallel to the K_d axis. In Fig. 5.5, the parametric region above $\mathscr{C}_1$ and to the right of $\mathscr{C}_2$ is stable, and the remaining regions are unstable. For example, for $\alpha = 1$, the points A_1 and A_2 are both stable but not A_3, while for $\alpha = 2$, the

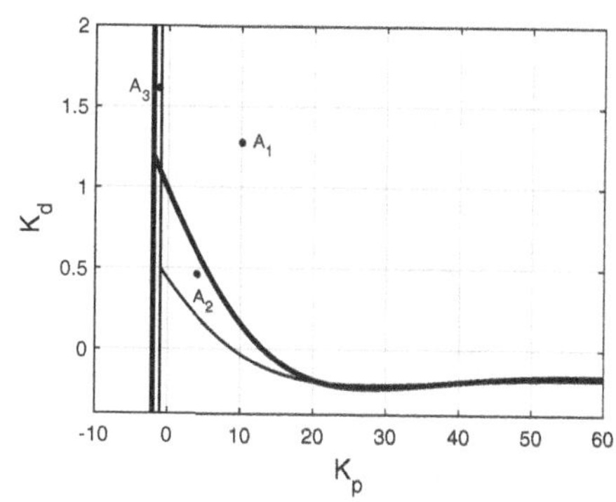

Fig. 5.5 Stability versus instability regions on the plane of controller gains K_p and K_d of the automated controller $u_k(t)$ in (5.19). Thin curves and thick curves correspond to, respectively, $\alpha = 1$ and $\alpha = 2$

[1] Note that the order of dynamics in (5.17) is different from that is (5.10); therefore, a simple matching between similar parametric settings is not possible.

points A_1 and A_3 are stable but not A_2. We note that the boundaries forming the stable regions have some degree of robustness as they appear qualitatively similar to those obtained for different numerical settings in [9].

Notice further that the last term in (5.22) is bounded, that is,

$$\left| \alpha e^{i\omega h} \frac{1 - e^{-i\omega\delta}}{i\omega\delta} \right| \leq \alpha, \quad \forall \omega, \tag{5.24}$$

and thus for sufficiently large K_p and K_d, it is possible to override the destabilizing effects of the delays, thereby guaranteeing the stability of (5.22).

5.3.3 The Case with γ-distribution with or Without Gap

We start with analyzing stability with respect to mean and variance and without gap in the γ-distribution. Given the system characteristic function for the problem with γ-distribution *without gap* and $p \in \{3, 4, \ldots, 10\}$, it is possible to check computationally whether or not the system is stable. To this end, an example with $n = 3$ vehicles and driver aggressiveness $\alpha_k = \alpha = 2.0$ is selected, and for a range of q, stability versus instability is calculated. One can then plot how q changes with respect to mean pq (Fig. 5.6-left) and variance pq^2 (Fig. 5.6-right). Notice in Fig. 5.6-left that the slopes of the lines are equal to p, and hence from this correspondence it is easy to match the p values to the plots in Fig. 5.6-right.

One key observation from Fig. 5.6 is the following. When the mean of the distribution increases, that is, the memory on average becomes older, then the corresponding variance also increases. Thus, older information in the feedback can still be handled without losing stability with a larger memory window but instability will be inevitable if the mean is larger than a threshold.

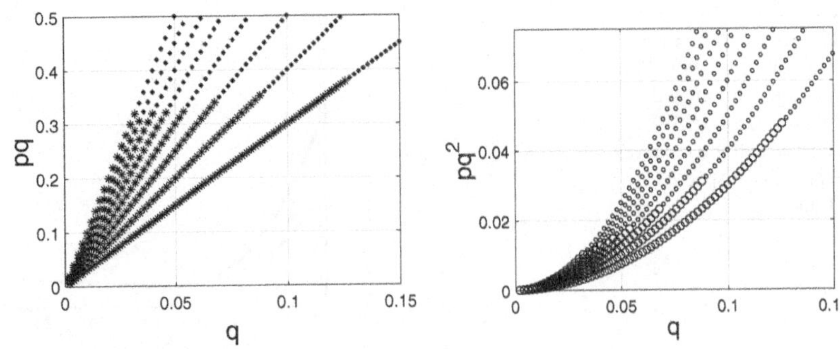

Fig. 5.6 Stability (larger markers) versus instability (smaller markers) in the parameter space of q versus mean pq (left) and variance pq^2 (right) for $p \in \{3, 4, \ldots, 10\}$. Slope of the lines on left figure increases with p. Ring configuration

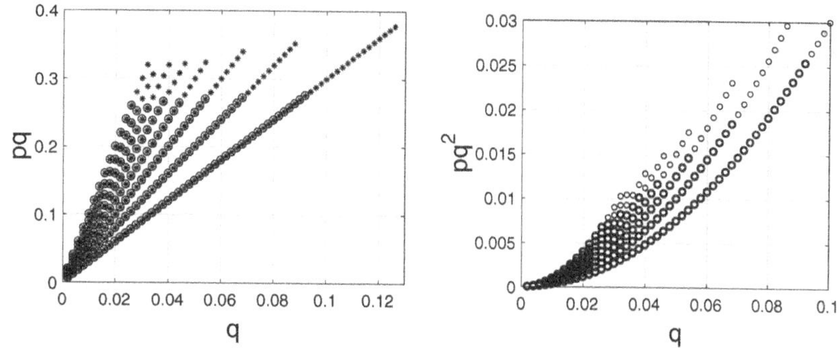

Fig. 5.7 Stable settings of the dynamics for $n = 3$ (smaller markers) and $n = 6$ (larger markers) on the parameter plane of q versus mean pq (left) and variance pq^2 (right) for $p \in \{3, 4, \ldots, 10\}$. Slope of the lines on left figure increases with respect to p. Ring configuration

We next study stability with respect to the number of vehicles, and without gap in the γ-distribution. In order to keep computational costs reasonable, here a scenario with $n = 6$ vehicles is considered, and compared with that in the previous case. For a concise presentation, only stability settings are plotted this time; see Fig. 5.7 where the $n = 3$ case is carried over from Fig. 5.6 for comparison. In terms of how the mean of the distribution changes, we see that when the number of vehicles is doubled, the maximum of the mean shrinks in the parameter space. Similarly, the range of variance is smaller for $n = 6$ as compared to $n = 3$.

The results indicate that when the number of vehicles is increased from three to six, in order to maintain stability, the average of the distribution cannot be as old and its spread cannot be as large. Last but not the least, both the three-vehicle and six-vehicle scenarios share common parameter settings where stability is maintained.

We conclude this subsection with a scenario including a γ-distribution with nonzero gap. To this end, we investigate how the gap h affects the stability of the dynamics. For this, the stability of the interconnection (5.10) is studied with respect to h in light of the discussions in Chap. 3. Following from the above results, here we pick two stable settings, one with $p = 3$ and the other with $p = 6$. The parameter q is fixed at $q = 0.05$.

These settings guarantee the stability of the system when the gap is zero, and moreover the mean of the distribution with gap is $h + pq$, where h must be selected less than a certain threshold to main stability. The analysis yields this threshold, i.e., the maximum allowable gap h, as shown in Fig. 5.8, with respect to the number of vehicles n in the ring configuration. Here, we find that as the number of vehicles in the ring configuration is increased, the maximum allowable delay gap h decreases, and is less for a larger p value, that is, for a larger mean and variance of the distribution. In other words, if the memory effect is pushed further into past and with sufficiently large spread, that is, when its mean and variance are increased then the amount of tolerable dead-time h must be decreased in order to maintain stability. Moreover, we see that, in the ring scenario with a large number of vehicles, the gap h converges to

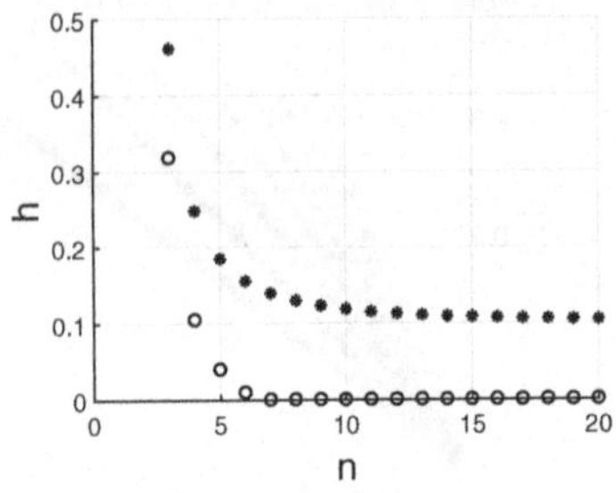

Fig. 5.8 Maximum allowable gap h while maintaining stability, plotted against the number of vehicles n in the ring configuration. Parameter values are $p = 3$ ($*$) and $p = 6$ ($\circ$); driver aggressiveness is set to $\alpha_k = 1.0$

0.1 for $p = 3$ and to 0^+ for $p = 6$. These quantities being much smaller than human reaction delays, we have that in both cases the drivers will not have sufficient gap time to react to stimuli, which could then destabilize the vehicle flow dynamics.

5.4 Notes and Further Reading

Some parts of the mathematical modeling and the pertaining discussions in this section are borrowed from [1, 9]. Moreover, human reaction delays in the context of traffic flow modeling and verification follow the arguments in [10, 11], as well as in [12, 13] regarding human physiology, anticipation, and memory effects. Effects of memory could indeed be influential on traffic flow dynamics as human drivers perform driving actions based on a collection of short-term past events.

A key message obtained from the case studies is that linear stability can hold in the presence of distributed delays, and if the distributions representing such delays are further pushed into past, then the distribution must be sufficiently spread over the history of the dynamics in order to maintain stability. In other words, stability can be achieved even with older information as long as there is sufficiently a large set of information, but there still exists a threshold for how old this information can be. This message consistently appears in several case studies and with different distribution functions considered. Beyond the application of memory modeling, distributed delays attracted much attention in the literature; see [14] in machine tool chatter modeling and [15] regarding how distributed delays can facilitate amplitude death among coupled oscillators. An investigation of the role of distributed delays in stability can be found in [16], where it is shown that distributed delays can improve stability over discrete delays.

A technical challenge in the stability analysis in the presence of distributed delays could be that some distributions could bring about the delay terms both in the exponents and as coefficients of the characteristic functions. Delay appearing also as a coefficient in the characteristic function can impose additional challenges in the stability analysis as demonstrated in example cases, mainly due to the fact that periodicity property of critical delays imposed by the trigonometric identities may no longer hold. Along these lines, for example, we refer the readers to a general treatment of a class of linear delay differential equations with delay coefficients [17]. Finally, a different frequency-sweeping method to treat the uniform distribution case can be found in [18]. For other topics, interested readers can refer to the following publications and the references therein [1, 19–22].

References

1. R. Sipahi, F.M. Atay, S.-I. Niculescu, Stability of traffic flow with distributed delays modeling the memory effects of the drivers. SIAM Appl. Math. **68**(3), 738–759 (2008)
2. F.M. Atay, On the duality between consensus problems and Markov processes, with application to delay systems. Markov Process. Relat. Field. **22**, 537–553 (2016)
3. R. Sipahi, F.M. Atay, S.-I. Niculescu, *Effects of Short-Term Memory of Drivers on Stability Interpretations of Traffic Flow Dynamics* (American Control Conference, New York City, 2007)
4. D. Swaroop, J.K. Hedrick, S.B. Choi, Direct adaptive longitudinal control of vehicle platoons. IEEE Trans. Veh. Technol. **50**(1), 150–161 (2001)
5. D. Swaroop, J.K. Hedrick, String stability of interconnected systems. IEEE Trans. Autom. Control **41**(3), 349–357 (1996)
6. P.A. Cook, Stable control of vehicle convoys for safety and comfort. IEEE Trans. Autom. Control **52**(3), 526–531 (2007)
7. J.E. Naranjo, C. Gonzalez, J. Reviejo, R. Garcia, T. de Pedro, Adaptive fuzzy control for inter-vehicle gap keeping. IEEE Int. Trans. Syst. **4**(3), 132–142 (2003)
8. P. Seiler, A. Pant, K. Hedrick, Disturbance propagation in vehicle strings. IEEE Trans. Autom. Control **49**(10), 1835–1842 (2004)
9. R. Sipahi, S.-I. Niculescu, Stability of Car Following with Human Memory Effects and Automatic Headway Compensation, theme issue Traffic jams—dynamics and control. Theme ed. by G. Stépán, G. Orosz, E. Wilson, *Philosophical Transactions of the Royal Society A*, vol. 368 (1928), pp. 4563–4583 (2010)
10. M. Bando, K. Hasebe, K. Nakanishi, A. Nakayama, Analysis of optimal velocity model with explicit delay. Phys. Rev. E **58**, 5429–5435 (1998)
11. A. Bose, P.A. Ioannou, Analysis of traffic flow with mixed manual and semi automated vehicles. IEEE Trans. Intell. Transp. Syst. **4**(4), 173–188 (2003)
12. M. Treiber, A. Kesting, D. Helbing, Delays, inaccuracies and anticipation in microscopic traffic models. Physica A **360**(1), 71–88 (2006)
13. M. Treiber, D. Helbing, Memory effects in microscopic traffic models and wide scattering in flow-density data. Phys. Rev. E **68**, 046119 (2003)
14. G. Stépán, *Retarded Dynamical Systems: Stability and Characteristic Function* (Longman Scientific & Technical, co-publisher Wiley, US, New York, 1989)
15. F.M. Atay, Distributed delays facilitate amplitude death of coupled oscillators. Phys. Rev. Lett. **91**, paper 094101 (2003)
16. F.M. Atay, Delayed feedback control near Hopf bifurcation. Discrete Contin. Dyn. Syst. - S **1**(2), 197–205 (2008)
17. E. Beretta, Y. Kuang, Geometric stability switch criteria in delay differential systems with delay dependent parameters. SIAM J. Math. Anal. **33**, 1144–1165 (2002)
18. X.-G. Li, S.-I. Niculescu, A. Cela, L. Zhang, Stability analysis of uniformly distributed delay systems: A frequency-sweeping approach, in *Delays and Interconnections: Methodology, Algorithms and Applications*, ed. by G. Valmorbida, A. Seuret, I. Boussaada, R. Sipahi, In Advances in Delays and Dynamics, vol. 10 (Springer Nature, Heidelberg, 2019), pp. 117–130
19. W. Qiao, R. Sipahi, A linear time-invariant consensus dynamics with homogeneous delays: analytical study and synthesis of rightmost eigenvalues. SIAM J. Control Optim. **51**(5), 3971–3992 (2013)

20. G. Orosz, Connected cruise control: modelling, delay effects, and nonlinear behaviour. Veh. Syst. Dyn. **54**(8), 1147–1176 (2016)
21. L. Zhang, G. Orosz, Motif-based design for connected vehicle systems in presence of heterogeneous connectivity structures and time delays. IEEE Trans. Intell. Transp. Syst. **17**(6), 1638–1651 (2016)
22. L. Zhang, J. Sun, G. Orosz, Hierarchical design of connected cruise control in the presence of information delays and uncertain vehicle dynamics. IEEE Trans. Control Syst. Technol. **26**(1), 139–150 (2017)

Chapter 6
Slinky Effects with Discrete and Distributed Delays

This chapter expands upon the discussions in Chaps. 4 and 5 by utilizing mathematical models with discrete or distributed delays to study slinky effects in traffic flow. As broadly discussed in the literature, slinky effects pertain to cyclic signals propagating in a group of vehicles, with growing amplitudes. In traffic flow, such effects are observed as cyclic perturbations in the headway, or velocities, emanating from upstream of the flow and eventually propagating downstream in the form of oscillations with growing amplitudes [1]. Larger headway amplitudes correspond to situations where headway alternates between too large and too small. When faced with smaller headways, drivers at the downstream of traffic have no choice but to stop their vehicles to prevent accidents. This behavior can in turn be reflected in increased traffic jams and reduced flow throughput.

While small yet growing amplitudes in headway propagation respect linear dynamics, as such amplitudes become relatively large, nonlinear effects will dominate the behavior of headway and ultimately jam formations. In this chapter, the aim is to put light on the mechanisms of amplitude amplifications only in the linear regime of the headway propagation dynamics. To this end, the analysis of slinky effects follows from frequency-response characteristics of linear traffic flow models. Moreover, this chapter specifically focuses on vehicles following each other in the linear (chain) configuration, but the provided guidelines can be used to extend to the case of ring configuration as well.

In the following, we start with some preliminaries to establish the fundamentals needed to study slinky effects. An illustrative example is provided next, and these tools are then implemented on models covered in Chaps. 4 and 5. One interest will be to investigate how slinky effects can be avoided under different delay models, driver aggressiveness parameters, and how such effects compare in the presence of constant-headway versus constant time-headway driving strategies.

R. Sipahi et al., *Deterministic Car-Following Traffic Models*,
SpringerBriefs in Control, Automation and Robotics,
https://doi.org/10.1007/978-3-031-58164-9_6

6.1 Preliminaries

For linear systems, slinky effects can be interpreted as the system becoming sensitive, e.g., around its resonance frequency, against some oscillatory perturbations. We will say that slinky effects arise when the amplitudes of the output oscillations are larger in steady state than those of the input oscillations, or equivalently when the ratio between the output amplitudes to the input amplitudes is greater than one. Notice that slinky effects and instability concepts are different. Instability can be avoided, or "asymptotic stability" can be attained, whenever a system is able to dissipate energy at a certain rate when subjected to "impulsive" inputs (see also Chap. 3). On the other hand, slinky effects can be avoided only if the output oscillation amplitudes are smaller than those of the input oscillations in steady state. An asymptotically stable system can very well exhibit slinky effects under certain conditions. However, if a system is unstable, then there is no need to study slinky effects because the system output will not remain bounded in steady state.

6.1.1 *Mathematical Framework*

Slinky effects in linear time-invariant systems can be studied using Fourier analysis, i.e., the frequency-response characteristics of the system. We first present the concept of slinky effects and tools for its analysis.

Consider a linear system S with input $u(t)$ and output $x(t)$. Let $G(s)$, $U(s)$, and $X(s)$ denote respectively the transfer function of this system, and the Laplace transforms of $u(t)$ and $x(t)$. Given $G(s)$, one should make sure first that the system being analyzed for slinky effects is asymptotically stable. In the case of non-delayed linear problems, this can be assessed using standard control theory tools. In the case of linear problems with delays, one may use the tools presented in Chap. 3 (see also the example cases in Chaps. 4 and 5). Stability analysis will reveal regions $\mathscr{P}$ in the parameter space where the system is asymptotically stable, for example, the range of delays for which stability holds.

Slinky effects are meaningful only for stable systems, that is, the study should be conducted in the stable parameter region $\mathscr{P}$. One then considers a periodic input to the system in the form $u(t) = \sin(\omega t)$, where ω is the excitation frequency. For a periodic input, a stable linear time-invariant system will produce, in steady state, a periodic output of the form $x(t) = M \sin(\omega t + \phi)$, where M and ϕ depend on ω, that is, $M = M(\omega)$ and $\phi = \phi(\omega)$; see Fig. 6.1. The question of interest is how M changes with respect to ω, as this quantity indicates whether or not the output amplitude is magnified against the input amplitude. For $M > 1$, we have slinky effects at the corresponding value of the input frequency.

To compute $M(\omega)$, a frequency-response analysis is adopted. To this end, one sets $s = i\omega$, $\omega \geq 0$, and studies the "magnitude" of the complex number $G(s)$, that is,

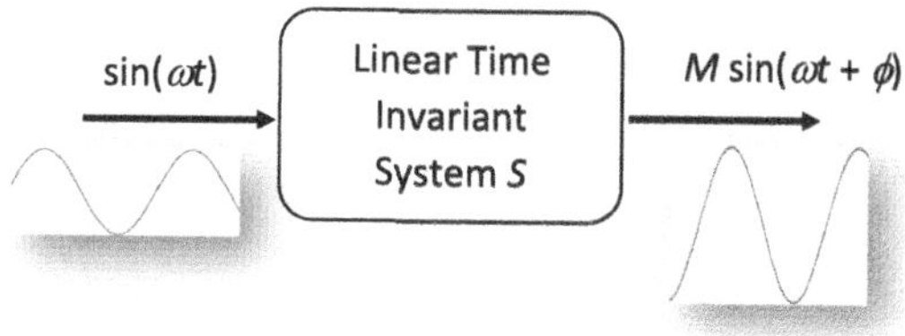

Fig. 6.1 The linear time-invariant system S is subject to a periodic input and produces a periodic output with amplitude M and phase ϕ. Here the case of $M > 1$ is depicted

$M(\omega) = |G(i\omega)|$. The magnitude $M(\omega)$ is indicative of how large the steady-state oscillations will be for a unit amplitude periodic input.

Remark 6.1 As slinky effects are related to the magnitude of the complex number $G(i\omega)$, we do not utilize the phase angle $\phi(\omega)$ between the input and output signals; however, for completeness we note that this phase angle is determined by the phase angle of the complex number, that is, $\phi(\omega) = \angle(G(i\omega))$.

6.1.2 *Illustrative Example*

Consider the transfer function

$$G(s) = \frac{1}{s^2 + 2s + 1 + 5e^{-\tau_1 s} + 4se^{-\tau_2 s}}. \tag{6.1}$$

As discussed above, the analysis for slinky effects must be performed in the parameter space $\mathscr{P}$ for which the dynamics is asymptotically stable, that is, for parameter values for which all roots of the characteristic equation $f := s^2 + 2s + 1 + 5e^{-\tau_1 s} + 4se^{-\tau_2 s} = 0$ have negative real parts. We recall that the stability of this system was assessed in Sect. 3.2.4.3 on the plane of τ_1–τ_2; thus, the stable parameter region $\mathscr{P}$ is the shaded region in Fig. 3.1. Given a set of delays in $\mathscr{P}$, we proceed to investigate the slinky effects by calculating the magnitude

$$M(\omega) := |G(i\omega)| = \frac{1}{\left|-\omega^2 + 2i\omega + 1 + 5e^{-i\omega\tau_1} + 4i\omega e^{-i\omega\tau_2}\right|}, \quad (\tau_1, \tau_2) \in \mathscr{P}.$$

For demonstration purposes, we pick three pairs of delay values from $\mathscr{P}$ and plot $M(\omega)$ against ω for each pair. The results are shown in Fig. 6.2. In this example, one pair of the selected delays leads to $|G(i\omega)| < 1, \forall\omega$ (thin curve), that is, the system is free of slinky effects. The other pairs of delays lead to $|G(i\omega)| > 1$ for some ω (star and circle markers); therefore, the system exhibits slinky effects at these delay settings. While the range of figure axes was adjusted for clarity, it is easy to show that $M(\omega) \to 0$ as $\omega \to \infty$. That is, for any delay selected from $\mathscr{P}$, there exists a frequency above which the magnitude M will be guaranteed to be less than unity.

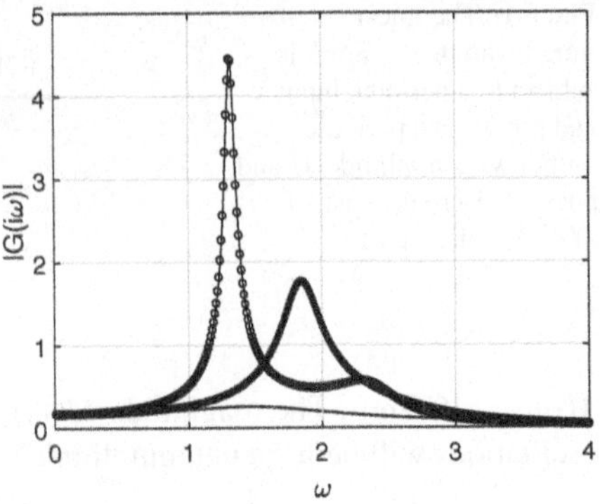

Fig. 6.2 Frequency-response magnitude of the dynamics in (6.1). Thin curve: $\tau_1 = 1.0$ and $\tau_2 = 0.4$; Stars: $\tau_1 = 1.5$ and $\tau_2 = 0.95$; Circles: $\tau_1 = 2.0$ and $\tau_2 = 1.2$. All delay pairs guarantee stability as they are selected from the shaded stability region $\mathscr{P}$ from Fig. 3.1

6.2 Slinky Effects Under Discrete Delays

We now turn our attention to traffic flow models with discrete delays, and implement the above presentation to such models. In the following, we study slinky effects for both constant-headway and constant time-headway driving strategies.

6.2.1 *Constant-Headway Model*

Consider the model

$$\dot{v}_k(t) = \alpha_k(v_{k+1}(t-\tau) - v_k(t-\tau)), \tag{6.2}$$

reviewed in Chap. 2, where $\alpha_k > 0$. Here, vehicle speeds relate to vehicle displacements as $\dot{x}_k(t) = v_k(t)$. In this setup, one obtains the following transfer function $G_k(s)$ between the velocities of two consecutive vehicles,

$$G_k(s) := \frac{\vartheta_k(s)}{\vartheta_{k+1}(s)} = \frac{\alpha_k s e^{-\tau s}}{s^2 + \alpha_k s e^{-\tau s}} = \frac{\alpha_k e^{-\tau s}}{s + \alpha_k e^{-\tau s}}, \tag{6.3}$$

where $\vartheta_k(s)$ is the Laplace transform of $v_k(t)$. Following [1] (see also Sect. 4.4.1), we let $D_k(s) = X_{k+1}(s) - X_k(s)$ denote the Laplace transform of the headway, which can be expressed as $D_k(s) = (\vartheta_{k+1}(s) - \vartheta_k(s))/s$. Noting that s in Laplace domain corresponds to the time-derivative operator, we have that velocity differences are given by $sD_k(s)$. Using (6.3), one can express $D_k(s)$ as $D_k(s) = (1 - G_k)\vartheta_{k+1}(s)/s$ and hence obtain the relationship between the dynamics of consecutive headways as

$$\frac{D_k(s)}{D_{k+1}(s)} = \frac{1 - G_k}{1 - G_{k+1}} G_{k+1} := \hat{G}_k(s). \tag{6.4}$$

It is then easy to see that the transfer function of consecutive relative velocities is given by the same expression $\vartheta_k(s)/\vartheta_{k+1}(s) = (sD_k(s))/(sD_{k+1}(s)) = \hat{G}_k(s)$. This function can be used to describe how headway and relative velocity propagate in a traffic flow scenario with n number of vehicles; notice here that the subscript runs up to $k = n - 2$. The transfer function of the headway propagation dynamics then reads

$$G_{q,n-1}(s) := \prod_{k=q}^{n-2} \hat{G}_k(s) = \frac{D_q(s)}{D_{q+1}(s)} \frac{D_{q+1}(s)}{D_{q+2}(s)} \cdots \frac{D_{n-2}(s)}{D_{n-1}(s)} \tag{6.5}$$

$$= \frac{D_q(s)}{D_{n-1}(s)} = \frac{1 - G_q(s)}{1 - G_{n-1}(s)} \prod_{k=q+1}^{n-2} G_k(s), \tag{6.6}$$

where $(n-2) \leq q \leq 1$.

The aim is to study whether, for any q, the transfer function $G_{q,n-1}(s)$ can amplify input headway amplitudes, i.e., induce slinky effects, with respect to certain frequencies in the periodic perturbations from D_{n-1}. Before studying slinky effects, one must make sure that the dynamics is stable. To this end, assume first that $\hat{G}_k(s)$, $k = 1, \ldots, n-1$, or equivalently $G_k(s)$, in this case, is a stable transfer function. Stability holds only for specific values of the pairs (τ, α_k). Consider that these values are collected in a set $\mathscr{P}$, that is, all the roots of the corresponding characteristic equation $f := s + \alpha_k e^{-s\tau} = 0$ are stable for $(\tau, \alpha_k) \in \mathscr{P}$ (see discussions on stability in Chap. 3). Then the effects of cyclic perturbations can be studied using $G_{q,n-1}(s)$ and for $(\tau, \alpha_k) \in \mathscr{P}$. For this purpose, we will adopt frequency-response methods to analyze the magnitude $|G_{q,n-1}(i\omega)|$ for $(\tau, \alpha_k) \in \mathscr{P}$. We say that slinky effects are avoided for all frequencies if and only if $\sup_\omega |G_{q,n-1}(i\omega)| < 1$, $\forall q$, which is equivalent to $\sup_\omega |\hat{G}_k(i\omega)| < 1$, $\forall k$.

Remark 6.2 For the car-following dynamics to be free of slinky effects, the conditions $\sup_\omega |\hat{G}_k(i\omega)| < 1$ should be separately checked for each $k = 1, 2, \ldots n-2$. For the sake of presentation, in the remainder of this chapter we demonstrate how to check this condition for a generic, fixed value of k. In this regard, the following discussions pertain to three vehicles, with a headway dynamics forming in between consecutive vehicles, and how this headway dynamics propagates from upstream to downstream. Once this is established, the reader can extend the study by considering all k values in a platoon of vehicles.

The condition $\sup_\omega |\hat{G}_k(i\omega)| < 1$ can be developed to better understand how the parameters τ and α_k could contribute to slinky effects. First of all, from (3.9), we know that the system is stable for $\tau \in [0, \tau_c)$, where $\tau_c = \pi/(2\alpha_k)$. Let $\tau < \tau_c$ hold and we study the magnitude of $\hat{G}_k$ given by

$$|\hat{G}_k(i\omega)| = \frac{\alpha_{k+1}}{\sqrt{\omega^2 - 2\omega\alpha_k \sin(\tau\omega) + \alpha_k^2}}. \tag{6.7}$$

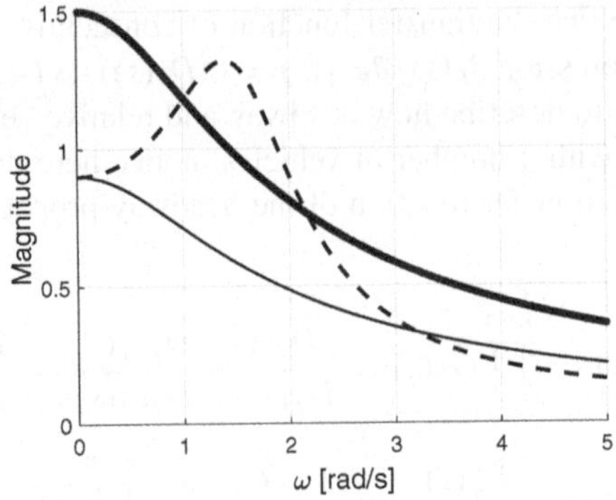

Fig. 6.3 Magnitude plot $|\hat{G}_k(i\omega)|$ with respect to ω. Case (i) (thick solid curve): $\alpha_1 = 1.0, \alpha_2 = 1.5$, $\tau = 0.2, \tau_c = \pi/2$. Slinky effects due to $\alpha_2/\alpha_1 > 1$. Case (ii) (thin solid curve): $\alpha_1 = 1.0, \alpha_2 = 0.9$, $\tau = 0.2, \tau_c = \pi/2$. No slinky effects due to relatively smaller delay $\tau < \tau_c$ even though $\alpha_2/\alpha_1 > 1$. Case (iii) (dashed curve): $\alpha_1 = 1.0, \alpha_2 = 0.9, \tau = 0.7854, \tau_c = \pi/2$. Slinky effects due to relatively larger delay τ even though $\alpha_2/\alpha_1 < 1$

Observe that $|\hat{G}_k(0)| = \alpha_{k+1}/\alpha_k$. Hence, if $\alpha_{k+1}/\alpha_k > 1$ then it is guaranteed that there exists a range of ω, $\omega \in [0, \overline{\omega}]$, for which slinky effects will occur. This simple rule indicates that if driver $(k + 1)$ is more aggressively regulating the speed of vehicle $(k + 1)$ than the follower driver k, slinky effects will arise. Thus, $\alpha_{k+1}/\alpha_k < 1$ is a necessary condition to prevent slinky effects.

We next present a numerical case study. We assume $\tau < \tau_c$ and consider three different scenarios, (i) slinky effects due to $\alpha_{k+1}/\alpha_k > 1$, (ii) no slinky effects due to relatively small delay τ and with $\alpha_{k+1}/\alpha_k < 1$, and (iii) slinky effects due to relatively large delay although $\alpha_{k+1}/\alpha_k < 1$. Figure 6.3 depicts the plot of (6.7) with respect to ω, displaying conformity with the above discussions. Firstly, even if human reaction delays are small, when the leader vehicle is more aggressive than the follower, slinky effects arise within a frequency band $\omega \in [0, \overline{\omega}]$ (here $\overline{\omega} \approx 1.5$ rad/s). When the aggressive driver is the follower, however, slinky effects can be avoided if human reaction delays are not very large. In the figure, we see this clearly with the magnitude decaying from 0.9 toward 0. On the other hand, with large human reaction delays, slinky effects will arise even if a more aggressive driver is the follower of a less aggressive driver.

Notice that, as per the stability analysis, the system is in transition from stability to instability when $\tau = \tau_c$. That is, the system has a pair of poles on the imaginary axis $s = i\omega$ and hence the characteristic function is zero, $f(i\omega; \tau_c) = 0$. Therefore, when a delay value in $\mathscr{P}$ approaches sufficiently close to the critical delay τ_c, from the left of τ_c, i.e., $\tau_c - \tau \to 0^+$, we have that $|f| \to 0$. Since $|f|$ appears in the denominator, compare (6.3) and (6.7), it is clear that the magnitude in (6.7) will grow without bound as $\tau \to \tau_c$. Hence, by continuity with respect to system parameters, there exists a critical delay value less than τ_c for which slinky effects are guaranteed to occur.

In Fig. 6.4, consistent with Remark 6.2, we present time simulations of headway variations as they propagate from between vehicles $k + 1$ and k to between vehi-

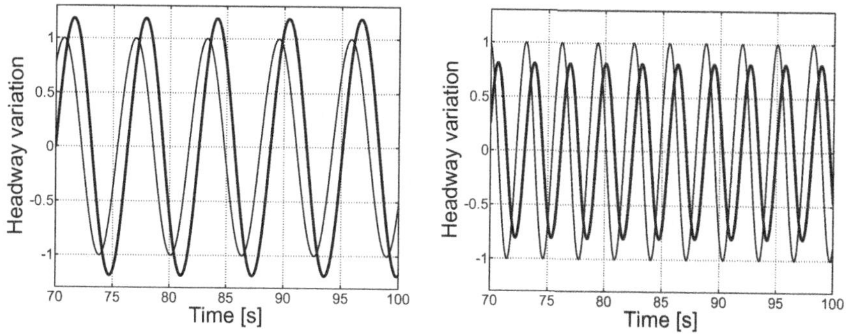

Fig. 6.4 Propagation of headway variation from the input (thin) to output (thick) at frequency $\omega = 1$ (left) and $\omega = 2$ (right) under constant-headway driving strategy with the settings in Case (i) of Fig. 6.3 ($\alpha_1 = 1.0, \alpha_2 = 1.5, \tau = 0.2$). Larger output headway variation on left (right) indicates slinky effects

cles k and $k - 1$. Since we are concerned with linear dynamics, we simulate only small headway variations, which should be interpreted as variations in the headway dynamics. To this end, and to emphasize the presence of slinky effects, we select the settings in Case (i) of Fig. 6.3 and simulate how the upstream headway variation at $\omega = 1$ and $\omega = 2$ rad/s propagates downstream. Since the frequency-response analysis pertaining to slinky effects are studied in steady state, the simulation results in Fig. 6.4 are shown after the transients die out and steady state is reached. Notice from Fig. 6.3 that when the excitation frequency of the upstream headway variation is at $\omega = 1$, the magnitude is approximately 1.2, which is consistent with the time simulations where we observe that downstream headway variations are about 20% larger than those of the input (left panel in Fig. 6.4); that is, the dynamics exhibits slinky effects. When the excitation frequency is increased to $\omega = 2$ however, the magnitude in Fig. 6.3 is about 0.8, which is consistent with the time simulations where we observe that downstream amplitudes attenuate at 20% in comparison to input amplitudes (right panel in Fig. 6.4).

6.2.2 Constant Time-Headway Model

We now refer to Sect. 4.4 where we discussed the asymptotic stability of a model with constant time-headway, in particular to Eq. (4.20). Notice in this model that h denotes the time-headway, and one recovers the discussions in Sect. 6.2.1 when $h = 0$. We now wish to investigate how h can influence slinky effects. In view of (4.20), we obtain $\hat{G}_k(s)$ as

$$\hat{G}_k(s) = \frac{\alpha_{k+1}(1 - h_k\alpha_k e^{-\tau s})e^{-\tau s}}{(1 - h_{k+1}\alpha_{k+1}e^{-\tau s})(s + \alpha_k e^{-\tau s})}, \tag{6.8}$$

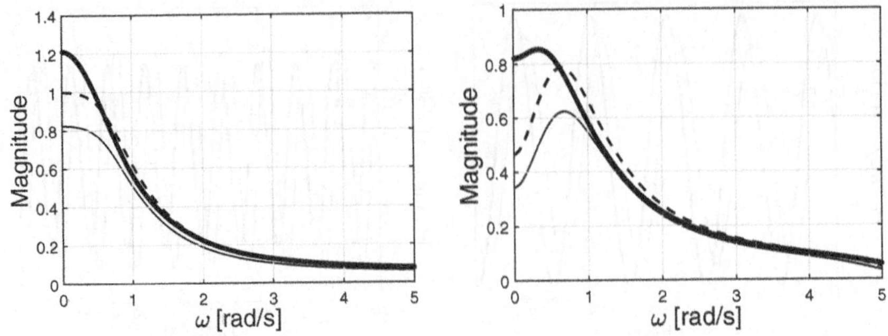

Fig. 6.5 Magnitude plot $|\hat{G}_k(i\omega)|$ with respect to ω. Left: constant-headway driving strategy $h_1 = h_2 = 0$. Right: constant time-headway driving strategy $h_1 = 1.8$ and $h_2 = 1.0$. Discrete delay $\tau = \tau_1 = \tau_2 = 1.2$ from Fig. 4.5. Case (i) (thick solid curve): $\alpha_1 = 0.33$, $\alpha_2 = 0.4$. Case (ii) (thin solid curve): $\alpha_1 = 0.4$, $\alpha_2 = 0.33$. Case (iii) (dashed curve): $\alpha_1 = 0.4$, $\alpha_2 = 0.4$

whose asymptotic stability was studied in Sect. 4.4.

To be consistent, the parameter settings are borrowed from Sect. 4.4 for α_k and α_{k+1}; see also Fig. 6.3 corresponding to constant-headway driving strategy. Following Sect. 4.4.1, notice that

$$\hat{G}_k(0) = \frac{\alpha_{k+1}}{\alpha_k} \frac{1 - h_k\alpha_k}{1 - h_{k+1}\alpha_{k+1}},$$

hence, the time-headway h_k can be expected to be useful for further reducing the magnitude of $\hat{G}_k(i\omega)$ at low frequencies. For example, when $\alpha_{k+1}/\alpha_k < 1$, which is a necessary condition to avoid slinky effects in the case of constant-headway driving strategy (see Sect. 6.2.1), and if one selects h_k relatively larger and h_{k+1} relatively smaller while maintaining $|h_k\alpha_k| < 1$ as part of the stability condition, it is possible with $h_k \neq 0$ that the magnitude at low frequencies can be made smaller. This is demonstrated in Fig. 6.5.

Some observations regarding Fig. 6.5 are in order, keeping in mind Remark 6.2. Firstly, when the drivers are not identical, with the aggressive one being followed by a less aggressive one (Fig. 6.5-left), we have from Sect. 6.2.1 that slinky effects arise under constant-headway driving strategy. On the other hand, when the drivers are identical (and comparably aggressive), or when the more aggressive one takes the follower role, then slinky effects are avoided under this driving strategy. Moreover, by incorporating constant time-headway in the driving strategy (Fig. 6.5-right), it becomes possible to further push the magnitude curve down, especially at low frequencies, with a proper selection of the time-headway. Slinky effects found in the case of constant headway by having the aggressive driver as the leader can be avoided without changing the driver aggressiveness but instead implementing the constant time-headway driving strategy.

We close this subsection by presenting time simulations of headway variations. Different from the previous subsection, we will compare the headway propagation dynamics between constant-headway and constant time-headway driving strategies,

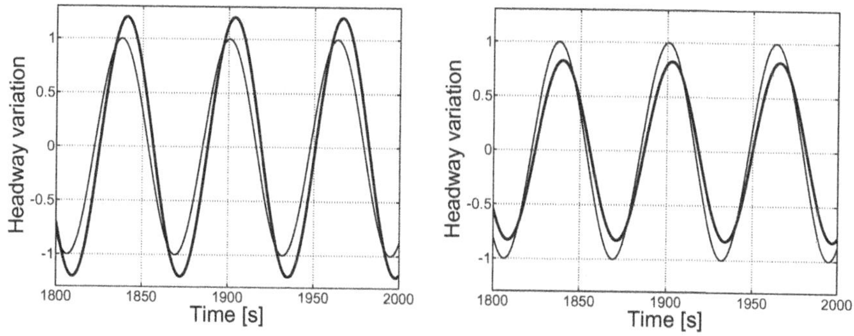

Fig. 6.6 Propagation of headway variation from the input (thin curve) to output (thick curve) at frequency $\omega = 0.1$ under constant-headway (left) and constant time-headway driving strategy (right) with the settings in Case (i) of Fig. 6.5. Larger output headway variation on left (right) indicates slinky effects

paralleling the results in Fig. 6.5. For demonstration purposes, we pick Case (i) in this figure, that is, $\alpha_1 = 0.33$, $\alpha_2 = 0.4$. Moreover, for both driving strategies we have $\tau = 1.2$, and in the case of constant-headway driving strategy, time headway is set to $h_1 = h_2 = 0$, and in the case of constant time-headway driving strategy we have $h_1 = 1.8$ and $h_2 = 1.0$. To demonstrate the benefits of utilizing constant time-headway driving strategy, we select the excitation frequency of the upstream headway variation to be at $\omega = 0.1$ rad/s. Simulation results in steady state are presented in Fig. 6.6.

Notice from Fig. 6.5 that when the excitation frequency of the upstream headway variation is at $\omega = 0.1$, the magnitude under constant driving strategy is about 1.2, which is consistent with the time simulations where we observe that downstream headway variations are about 20% larger than those of the input (left panel in Fig. 6.6); that is, the dynamics exhibits slinky effects. When constant time-headway driving strategy is utilized instead, then the magnitude in Fig. 6.5 is about 0.8, which is consistent with time simulations where we observe that downstream amplitudes attenuate at 20% in comparison to input amplitudes (right panel in Fig. 6.6).

6.3 Slinky Effects Under Distributed Delays

We now study slinky effects for traffic flow models with distributed delays, following a similar presentation layout as for discrete delays. We recall once again that the discussions below should be viewed in light of Remark 6.2.

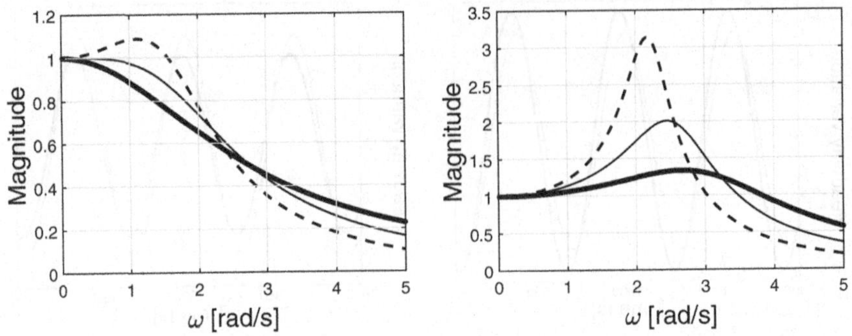

Fig. 6.7 Magnitude plot $|\hat{G}_k(i\omega)|$ with respect to ω considering drivers' uniformly distributed delays. Left: low gain $\alpha_k = 1.0$. Right: high gain $\alpha_k = 2.0$. Case (i) (thick solid curve): $h = 0.25$, $\delta = 0.25$. Case (ii) (thin solid curve): $h = 0.25$, $\delta = 0.5$. Case (iii) (dashed curve): $h = 0.25$, $\delta = 0.75$

6.3.1 Constant-Headway Model

We start with the constant-headway model. Specifically, we utilize the model (5.1) with uniform distribution whose transfer function representation is given in (5.7). In view of (6.4), we have

$$\hat{G}_k(s) = \frac{\alpha_{k+1} e^{-sh}(1 - e^{-s\delta})}{s^2\delta + \alpha_k e^{-sh}(1 - e^{-s\delta})}, \tag{6.9}$$

where the parameters h and δ of the uniform distribution (5.4) and the driver aggressiveness α_k are to be selected from the parameter space $\mathscr{P}$ guaranteeing asymptotic stability. For example, from Fig. 5.1 under linear configuration of vehicles, we can investigate which pairs (h, δ) can help avoid slinky effects. Moreover, for a weaker gain, the stability region in Fig. 5.2 is much larger, and one may wonder how weaker or stronger gains influence slinky effects. To this end, we investigate $|\hat{G}_k(s)|$ for various choices of h and δ for both low gain $\alpha_k = 1.0$ and high gain $\alpha_k = 2.0$ driver aggressiveness constants. Moreover, as a case study, we fix the gap time $h = 0.25$ and study the influence of the memory window δ on slinky effects. The results are presented in Fig. 6.7, where we observe that the low gain case (Fig. 6.7-left) is much less sensitive to periodic excitations, but, even in this case, the system remains free of slinky effects only for relatively small memory window size δ. With increased driver aggressiveness, slinky effects become much more pronounced (Fig. 6.7-right).

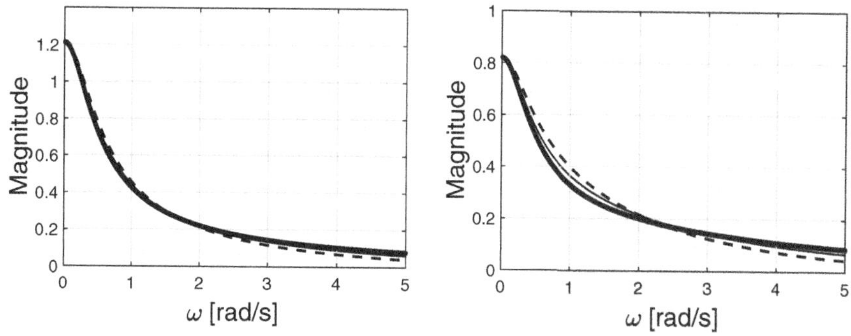

Fig. 6.8 Magnitude plot $|\hat{G}_k(i\omega)|$ with respect to ω considering drivers' uniformly distributed delays. A relatively more aggressive driver leads a less aggressive driver: $\alpha_k = 0.33$ and $\alpha_{k+1} = 0.4$. Left: constant-headway driving strategy $h_1 = h_2 = 0$. Right: constant time-headway driving strategy $h_1 = 1.8$ and $h_2 = 1.0$. Case (i) (thick solid curve): $h = 0.25$, $\delta = 0.25$. Case (ii) (thin solid curve): $h = 0.25$, $\delta = 0.5$. Case (iii) (dashed curve): $h = 0.25$, $\delta = 0.75$

6.3.2 Constant Time-Headway Model

In a similar way, one can incorporate a constant time-headway driving strategy into the distributed delay model, following the footprints in the discrete delay case. While the formulation is straightforward, the arising transfer function $\hat{G}_k(s)$ is lengthy and hence omitted here. Moreover, with the introduction of the time-headway parameter h_k, a necessary condition for the stability of the arising dynamics is given by $|\alpha_k h_k| < 1$, see [2]. In the example case provided next, this condition is respected, and a set of parameters consistent with those in Fig. 6.7 are utilized for comparison purposes. Results with constant-headway versus constant time-headway driving strategies are presented in Fig. 6.8. In this study, we have assumed that a more aggressive driver is in front of a relatively less aggressive driver, $\alpha_{k+1} > \alpha_k$. In such cases, constant-headway driving strategy is not desirable (Fig. 6.8-left) since slinky effects arise at low frequencies. Clearly, introducing the constant time-headway $h_k \neq 0$ helps push the entire magnitude curve below unity, thus eliminating the slinky effects (Fig. 6.8-right).

Remark 6.3 The above discussions are presented in view of Remark 6.2 hence they are only valid for three vehicles, forming two headway dynamics each between consecutive vehicles, and how slinky effects are characterized based on these two headway dynamics. These results can be extended to the case of more than three vehicles. For example, if one would be interested in studying how more aggressive drivers are leading less aggressive drivers in a platoon of vehicles, this would then require distributing α_k values such that $\alpha_{k+1} > \alpha_k$ with $k = 1, 2, \ldots n - 1$, and studying the headway propagation dynamics for $k = 1, 2, \ldots n - 2$.

6.4 Notes and Further Reading

This chapter focused on understanding how periodic perturbations in the headway of upstream traffic can propagate downstream and influence the downstream headway. We have combined the stability analysis from Chaps. 4 and 5 with frequency-response analysis to investigate the characteristics of this propagation. The results indicate either a propagation pattern that amplifies in downstream direction, which is termed slinky effects, or a pattern in which headway attenuates toward downstream, indicating that the flow is free of slinky effects.

While slinky effects are not technically considered as "instability", the presence of slinky effects is undesirable as it may lead to undesirable headway dynamics, either too large or too small. When headway becomes too small, this can cause drivers to slow, or even stop their vehicles, triggering jam formations, which shows the importance of slinky effects.

In this chapter, we have studied slinky effects under two distinct driving strategies, in the presence of either discrete or distributed delays. Under each strategy we considered vehicles' speeds are regulated as a function of constant headway between the vehicles, or as a function of how fast they are traveling, as quantified by time-headway.

Several interesting results are obtained in the analysis of slinky effects. First of all, a more aggressive driver being the leader of a less aggressive driver induces slinky effects. Further, larger human reaction delays also invite slinky effects, and in general there exists a critical delay value larger than which slinky effects cannot be avoided. Another critical result is that time-headway is quite useful in eliminating slinky effects in a comparable case study, although it is subject to a stricter stability condition. This suggests that a driving strategy based on time-headway, or equivalently, based on some anticipation characteristics, has the potential to render a traffic flow free of slinky effects and potentially help reduce jam formations.

Interested readers are referred to the following studies and the references therein to further expand on this important topic. For the fundamental derivations, see [3, 4] as well as [1] where the presence of human reaction delay is considered. For a study of slinky effects under human reaction delays, see [5] with discrete delays and [2] with distributed delays. For a characterization of jam-free versus jammed traffic flow with the consideration of human anticipation, see [6]. For model-based prediction of jam formations see [7, 8]. Finally, for the design and analysis of vehicle motifs, the readers are referred to [9, 10].

References

1. A. Bose, P.A. Ioannou, Analysis of traffic flow with mixed manual and semi automated vehicles. IEEE Trans. Intell. Transp. Syst. **4**(4), 173–188 (2003)
2. R. Sipahi, F.M. Atay, S.-I. Niculescu, Stability analysis of a constant time-headway driving strategy with driver memory effects modeled with distributed delays, in *IFAC Workshop on Time Delay Systems,* Ann Arbor, MI (2015)
3. D. Swaroop, J.K. Hedrick, String stability of interconnected systems. IEEE Trans. Autom. Control **41**(3), 349–357 (1996)
4. D. Swaroop, J.K. Hedrick, S.B. Choi, Direct adaptive longitudinal control of vehicle platoons. IEEE Trans. Veh. Technol. **50**(1), 150–161 (2001)
5. R. Sipahi, S.-I. Niculescu, Chain stability in traffic flow with driver reaction delays, *American Control Conference, Seattle, WA* (2008)
6. M. Treiber, A. Kesting, D. Helbing, Delays, inaccuracies and anticipation in microscopic traffic models. Physica A **360**(1), 71–88 (2006)
7. G. Orosz, B. Krauskopf, R.E. Wilson, Bifurcations and multiple traffic jams in a car-following model with reaction time delay. Phys. D **211**, 277–293 (2005)
8. G. Orosz, R.E. Wilson, R. Szalai, G. Stépán, Exciting traffic jams: nonlinear phenomena behind traffic jam formation on highways. Phys. Rev. E **80**, Article 046205 (2009)
9. L. Zhang, G. Orosz, Motif-based design for connected vehicle systems in presence of heterogeneous connectivity structures and time delays. IEEE Trans. Intell. Transp. Syst. **17**(6), 1638–1651 (2016)
10. L. Zhang, J. Sun, G. Orosz, Hierarchical design of connected cruise control in the presence of information delays and uncertain vehicle dynamics. IEEE Trans. Control Syst. Technol. **26**(1), 139–150 (2017)

Series Editor Biographies

Tamer Başar is with the University of Illinois at Urbana-Champaign, where he holds the academic positions of Swanlund Endowed Chair, Center for Advanced Study (CAS) Professor of Electrical and Computer Engineering, Professor at the Coordinated Science Laboratory, Professor at the Information Trust Institute, and Affiliate Professor of Mechanical Science and Engineering. He is also the Director of the Center for Advanced Study-a position he has been holding since 2014. At Illinois, he has also served as Interim Dean of Engineering (2018) and Interim Director of the Beckman Institute for Advanced Science and Technology (2008–2010). He received the B.S.E.E. degree from Robert College, Istanbul, and theM.S., M.Phil., and Ph.D. degrees from Yale University. He has published extensively in systems, control, communications, networks, optimization, learning, and dynamic games, including books on non-cooperative dynamic game theory, robust control, network security, wireless and communication networks, and stochastic networks, and has current research interests that address fundamental issues in these areas along with applications in multi-agent systems, energy systems, social networks, cyber-physical systems, and pricing in networks.

In addition to his editorial involvement with these Briefs, Başar is also the Editor of two Birkhäuser series on Systems and Control: Foundations and Applications and Static and Dynamic Game Theory: Foundations and Applications, the Managing Editor of the Annals of the International Society of Dynamic Games (ISDG), and member of editorial and advisory boards of several international journals in control, wireless networks, and applied mathematics. Notably, he was also the Editor-in-Chief of Automatica between 2004 and 2014. He has received several awards and recognitions over the years, among which are theMedal of Science of Turkey (1993); Bode Lecture Prize (2004) of IEEE CSS; Quazza Medal (2005) of IFAC; Bellman Control Heritage Award (2006) of AACC; Isaacs Award (2010) of ISDG; Control Systems Technical Field Award of IEEE (2014); and a number of international honorary doctorates and professorships. He is a member of the US National Academy of Engineering, a Life Fellow of IEEE, Fellow of IFAC, and Fellow of SIAM. He has served as an IFAC Advisor (2017–), a Council Member of IFAC (2011–2014),

R. Sipahi et al., *Deterministic Car-Following Traffic Models*,
SpringerBriefs in Control, Automation and Robotics,
https://doi.org/10.1007/978-3-031-58164-9

President of AACC (2010–2011), President of CSS (2000), and Founding President of ISDG (1990–1994).

Miroslav Krstic is Distinguished Professor of Mechanical and Aerospace Engineering, holds the Alspach Endowed Chair, and is the Founding Director of the Cymer Center for Control Systems and Dynamics at UC San Diego. He also serves as Senior Associate Vice Chancellor for Research at UCSD. As a graduate student, he won the UC Santa Barbara best dissertation award and student best paper awards at CDC and ACC. He has been elected as Fellow of IEEE, IFAC, ASME, SIAM, AAAS, IET (UK), AIAA (Assoc. Fellow), and as a foreign member of the Serbian Academy of Sciences and Arts and of the Academy of Engineering of Serbia. He has received the SIAM Reid Prize, ASME Oldenburger Medal, Nyquist Lecture Prize, Paynter Outstanding Investigator Award, Ragazzini Education Award, IFAC Nonlinear Control Systems Award, Chestnut textbook prize, Control Systems Society Distinguished Member Award, the PECASE, NSF Career, and ONR Young Investigator awards, the Schuck ('96 and '19) and Axelby paper prizes, and the first UCSD Research Award given to an engineer. He has also been awarded the Springer Visiting Professorship at UC Berkeley, the Distinguished Visiting Fellowship of the Royal Academy of Engineering, and the Invitation Fellowship of the Japan Society for the Promotion of Science. He serves as Editor-in-Chief of Systems and Control Letters and has been serving as Senior Editor for Automatica and IEEE Transactions on Automatic Control, as Editor of two Springer book series-Communications and Control Engineering and SpringerBriefs in Control, Automation and Robotics-and has served as Vice President for Technical Activities of the IEEE Control Systems Society and as Chair of the IEEE CSS Fellow Committee. He has coauthored 13 books on adaptive, nonlinear, and stochastic control, extremum seeking, control of PDE systems including turbulent flows, and control of delay systems.

Index

R. Sipahi et al., *Deterministic Car-Following Traffic Models*,
SpringerBriefs in Control, Automation and Robotics,
https://doi.org/10.1007/978-3-031-58164-9

SPRINGER NATURE

GPSR Compliance

The European Union's (EU) General Product Safety Regulation (GPSR) is a set of rules that requires consumer products to be safe and our obligations to ensure this.

If you have any concerns about our products, you can contact us on ProductSafety@springernature.com

In case Publisher is established outside the EU, the EU authorized representative is:

Springer Nature Customer Service Center GmbH
Europaplatz 3
69115 Heidelberg, Germany

The manufacturer's authorised representative in the EU is Springer Nature Customer Service Centre GmbH, Europaplatz 3, 69115 Heidelberg, Germany. If you have any concerns regarding our products, please contact ProductSafety@springernature.com

Printed and bound by CPI Group (UK) Ltd, Croydon, CR0 4YY
28/09/2026
02224073-0013